THE INSTRUCTOR'S EDITION OF

# Fish Farm

## A Simulation of Commercial Aquaculture

Robert Kosinski
Clemson University

**The Benjamin/Cummings Publishing Company, Inc.**
Redwood City, California • Menlo Park, California
Reading, Massachusetts • New York • Don Mills, Ontario
Wokingham, U.K. • Amsterdam • Bonn • Sydney
Singapore • Tokyo • Madrid • San Juan

*Sponsoring Editor:* Edith Beard Brady
*Editorial Coordinator:* Valerie Kuletz
*Production Supervisor:* Larry Olsen
*Copyeditor:* Eleanor Brown
*Artist:* John Norton
*Designer:* Vicki Philp

*The cover:* Computer-generated images of four important aquacultural fish (carp, channel catfish, rainbow trout, and Tilapia) superimposed on a graph showing mean profit and range of profit from a series of *Fish Farm* stocking density experiments (see page 54 of the student manual).



ISBN 0-8053-1898-4

1 2 3 4 5 6 7 8 9 10--AL--97 96 95 94 93

**The Benjamin/Cummings Publishing Company, Inc.**
390 Bridge Parkway
Redwood City, California 94065

# PREFACE for the Instructor's Guide

Although training in the scientific method is one of the major purposes of laboratory courses, it is often hard to accomplish with traditional exercises. In order to cover a large amount of content review, usually in coordination with a lecture course, lab courses tend to become collections of short, "sure-fire" experiments which reliably demonstrate basic principles and whose correct outcomes are well known in advance. Because of budget, logistic, and time constraints, the student is not given an opportunity to devise novel experiments or interpretations. The view of science as a compilation of facts rather than as a method of mental operation is reinforced by many traditional laboratory programs.

Fully investigative labs, in which content coverage is much less important and the student devises his or her own experiments, provide more training in the scientific method. At Clemson University, we have had fully investigative labs in the first semester of our non-majors biology course since early 1989, and it is obvious that this type of lab program is logistically challenging and makes heavy demands on instructors and students.

***Fish Farm's* role.** *Fish Farm* was designed as an introduction to the scientific method in the laboratory component of our introductory biology course, but it can be easily incorporated into a traditional course and used equally well with biology majors and non-majors.

*Fish Farm* is a computer simulation (available for Apple® II, Macintosh® and IBM® computers) of an aquaculture enterprise. The student attempts to experimentally determine the economically optimum culturing conditions for a new type of fish. After these optimum conditions have been determined, success is assessed by the profits of five years of simulated commercial operation. If you are interested in introducing investigative elements into a traditional lab course, or if you want to to offer a more complex challenge to the students in an investigative lab course, *Fish Farm* has several advantages:

1. *Fish Farm* is a simulation. If computers are available, it has minimal logistic requirements. Experiments which are long, costly, and technically difficult can be simulated in a matter of minutes. The two major barriers in "wet" labs--lack of time and lack of student skill--are overcome.

2. The overall objective of profitable aquaculture is easily understood by the students, and assessing success is easy. Competition for the highest profits will add motivation.

3. The problem demands a rigorous scientific approach. There are five culturing conditions which must be at the right value simultaneously, so guessing and random shifts in conditions are not usually successful. On the other hand, a systematic series of experiments will allow most students to solve the problem.

4. The results of each experiment contribute to the solution of the remaining ones.

5. Experiments performed under identical conditions do not have identical outcomes. The student learns the importance of replicate measurements to counter variability.

6. The correct answer is *not* known in advance. There are 26 "unknown" fish with a wide range of optimal culturing conditions (plus channel catfish, whose culturing conditions are known). The only way to determine the correct culturing conditions for an unknown fish is to consult experimental data.

*Fish Farm* is a *stand-alone* exercise: the student manual contains all the content needed for successful completion. However, *Fish Farm* is not a *do-alone* exercise. It should be used with groups of students and with instructor support. At Clemson, we have used it during the first two meetings of an investigative laboratory course.

This manual consists of an *Instructor's Guide* which is interleaved with the student manual. Each chapter of the *Instructor's Guide* comes before the corresponding student manual material. An instructor planning to use *Fish Farm* should read the student manual because it contains all the aquaculture background and procedures; this *Instructor's Guide* confines itself to practical advice for those supervising the use of *Fish Farm*.

The *Instructor's Guide* has four chapters, and its organization follows the two-laboratory method of using *Fish Farm*. Chapter 1 introduces *Fish Farm* and gives the instructor practical information such as how much time to allot to the exercises. Chapter 2 gives step-by-step instructor procedures for running the first *Fish Farm* lab, and Chapter 3 does the same thing for the second *Fish Farm* lab. Chapter 4 gives tips on grading the *Fish Farm* report. Finally, Appendices A-D give software installation information, transparency masters and suggestions for using *Fish Farm* for more purely ecological types of exercises.

**Acknowledgments.** The Fund for the Improvement of Postsecondary Education, U.S. Department of Education, supported the development of *Fish Farm* in 1987 as part of a larger investigative laboratory project (Award GOO8730323). The National Science Foundation (NSF) funded purchase of computers for use of *Fish Farm* at Clemson (Award DIR-8852751), and later the NSF funded a national workshop on our investigative laboratory program (Award USE-8954406).

Without this funding there would have been no *Fish Farm*, but funded activities were only part of the story. *Fish Farm* has been used every semester at Clemson University since spring, 1989, and the program has constantly improved because so many people have taken the time and trouble to comment on it. My first thanks go to my programmers, Ted Schindler (IBM version) and Rodney Cummings (Apple II and Macintosh versions). I am not an aquaculturist, but Thomas Schwedler of the the Clemson Aquaculture, Fisheries, and Wildlife Biology Department kept me firmly aligned with aquacultural reality during program development. I got many helpful suggestions about the best way to use *Fish Farm* in class from my faculty colleagues Jean Dickey, Kenyon Revis-Wagner, and John Cummings, plus more suggestions from many Clemson teaching assistants, especially Darrell Bayles. Jean Dickey and I held a 1990 investigative lab workshop which featured *Fish Farm*, and I thank all workshop participants for their suggestions, especially Richard Swade of California State University at Northridge. Dean Decker of the University of Richmond and Eugene Kaplan of Hofstra University are other loyal *Fish Farm* users who gave me many helpful ideas. The detailed fish drawings and the delightful cartoons which adorn the student manual are the work of John Norton, a talented and versatile biological illustrator based in Annapolis, Maryland.

Lastly, I'd like to dedicate this work to my wife, Margaret, and to my children, Bobby and Jody. My family grew to dread the phrase "I have to work on *Fish Farm*," but nevertheless accepted these lengthy periods with good humor.

Robert J. Kosinski
Clemson University

# CONTENTS for the Instructor's Guide

INSTRUCTOR CHAPTER 1

# Introduction to *Fish Farm*

If you want to get started *right away*, directions for installing *Fish Farm* on your computer appear in Appendix A (page I-30 of this *Instructor's Guide*).

*Fish Farm* is a computer simulation of an aquaculture enterprise in which the student attempts to experimentally determine the economically optimum culturing conditions for a new type of fish. After these optimum conditions have been determined, success is assessed by the profits of five years of simulated commercial operation. Students who have experimented systematically and have drawn the correct conclusions will make about $250,000 per growing season from their simulated farm. If a student has a defective experimental program, the farm will suffer from oxygen depletion, ammonia toxicity, epidemics and spectacular fish kills which may cost millions of dollars.

With a minimum of logistical support, the *Fish Farm* problem allows students to solve an complex, multifactor problem whose correct answer is *not* known in advance. Students working on *Fish Farm* are given a structure and an intermediate level of direction (e.g., they are told they must perform a feed protein experiment and they are told how to do it, but they are not told which feed protein contents to try, and determining when to stop is their decision).

## The *Fish Farm* Exercises

In *Fish Farm*, the student plays the role of the head of the research division of a aquaculture company which specializes in catfish "growout" (raising catfish from the fingerling stage to about 1 pound). But now the company has acquired a new variety of fish. The company plans to raise this fish in a production unit near Charleston, South Carolina; the unit has 200 hectares (about 495 acres) of 1 m deep ponds.

The student must first determine the tolerance of the fish to temperature extremes. If the fish can tolerate summer pond temperatures, then it can be raised without groundwater input into the pond. But if the fish must be kept cooler than these temperatures, it will have to be raised at high density in a small pond area with a rapid water flowthrough.

Next, when oxygen gets low in aquaculture ponds, the fish can be saved by aeration of the water with a paddlewheel. For most economical use of the aerator, the student must determine the dissolved oxygen concentration at which the fish first become stressed. This is the dissolved oxygen concentration at which the aerator will come on.

The student then must determine the crude protein content of the feed which results in the cheapest weight gain.

The next experimental step is to determine stocking density, or the number of fish to put in each pond, and when to harvest the fish to maximize profit.

Finally, the student tests the recommended culturing conditions with a "production run" simulating five years of commercial operation.

## The Schedule of *Fish Farm* Events

Each institution may have a different way of using *Fish Farm*. This manual describes only the way we use it at Clemson. At Clemson, *Fish Farm* occupies the first two lab meetings of an investigative lab course which meets three hours per week. The major events and requirements for the students under this regime are:

**Before Lab 1:** Read Chapters 1 and 2 of the *Fish Farm* student manual.

**Lab 1:** Orientation exercise and "tank" experiments. Students explore the operation of the program and test their ability to raise catfish profitably. Worksheet 1 (page 20 of the student manual) is completed in class. An "unknown" fish is assigned to each team of 3 or 4 students. The team will determine this fish's response to temperature, oxygen and feed protein.

**Homework:** Students read Chapters 3 and 4 of the *Fish Farm* manual. *Each student* does Worksheet 2 and Worksheet 3 (pages 49 and 65 of the student manual). Worksheet 2 will use the *Fish Farm* data collected during Lab 1.

**Lab 2:** Each team determines harvest time and optimum stocking rate for their fish. Using this information, each team performs the production run.

**Homework:** *Each student* writes a final report (directions in Chapter 5 of the student manual).

## The *Fish Farm* Report

The report, which is described in Chapter 5 of the student manual, consists of four basic parts: a cover, a raw data section (which consists of the data pages on pages 23, 27, 30, 67, and 75 of the student manual), a results section (containing formal tables and graphs on sheets of graph paper), and a discussion/conclusions section.

This report format differs from the standard sequence of introduction, materials and methods, results and discussion. The primary reason that this shortened report is used at Clemson is to reduce the workload on the students in an investigative course which has several oral and written reports in addition to *Fish Farm*. Also, since *Fish Farm* is a simulation exercise, there are less compelling reasons to include an introduction and a materials and methods section (especially the latter, since it could only describe operation of the program). Good data recording skills are encouraged by checking on the students' raw data, so raw data pages are made a formal part of the report.

## *Fish Farm* Logistics

The two basic ways of using *Fish Farm* are as an in-class exercise and as an exercise that the students complete on their own outside of class. At Clemson it is used in class, and this approach is recommended for other users, mainly because instructor explanations can quickly get the students over problems. Students using the program outside of class often run into difficulties because of the widespread tendency of lab students to skip right to the instructions without reading the background material. However, if the students will read the manual, or if instructor support is readily available, students can use *Fish Farm* quite successfully outside of class. All the information students need to use *Fish Farm* is contained in the student manual.

**If *Fish Farm* is used in class,** you should have one computer for each team of 3-4 students, and you should assign six hours of class time to exercises and explanations.

**If *Fish Farm* is used outside class,** try to assign pairs of students to work together. This will not only reduce the need for computers, but will also make the exercise a venture in cooperative problem solving. Each of these teams will need access to a computer for about 5 hours. Instructor support should be available (during office hours if not right at the point of use).

## Description of the *Fish Farm* Program

*Fish Farm* contains two types of experiments: tank and pond.

### Tank Experiments

In the tank experiments, 10 fish are placed in an aquarium with rapid water flowthrough for 40 days. The user can set the temperature, oxygen and feeding conditions. Because of the rapid water replacement, the fish cannot degrade their environment and so the fish are responding only to the conditions set by the user. Using the Clemson schedule in which *Fish Farm* is done in two lab periods, the tank experiments are completed during the first period.

### Pond Experiments

The pond experiments are conducted in 1 m deep outdoor ponds near Charleston, South Carolina. Conditions here vary dramatically over the 225 day growing season from March 21 through November 5. The ponds experience fluctuations of temperature and changes in the concentrations of oxygen, ammonia, algal nutrients, and disease-causing microbes. If conditions are degraded too severely, the fish go off their feed, get sick, and die. Occasionally, there will be spectacular fish kills and the farm will lose millions of dollars. On the other hand, if conditions are good, the fish will grow smoothly and the farm will make about $250,000 per growing season. If any fish die, a "Pathologist's Report" will give a breakdown of the causes of death (which include cold, heat, temperature change, starvation, ammonia poisoning, disease, and fighting). The two pond experiments are the stocking density experiment and the production run, and they are both completed during the second *Fish Farm* lab.

*Fish Farm* shows the student either digital or graphic displays. It also has a database for storing the results of past experiments.

## The Digital Displays

The digital displays show the numerical values of pond or tank temperature, oxygen, feeding rate, feed given, fish weight, number of surviving fish, and feed conversion ratio (a measure of fish assimilation efficiency). If a pond experiment is being done, the biomass of fish per hectare and the profits and losses are added to this list.

## The Graphic Displays

The graphic displays show either graphs of conditions in the current experiment or graphs of the results of past experiments. The current experiment graphs show temperature, dissolved oxygen, average fish weight, and profits of the experiment in progress. Because of memory limitations, the Apple® II version uses low-resolution graphics and only presents the last 20 days of data. After an experiment is over, the Macintosh® and IBM® versions (but not the Apple II version) allow the student to review these graphs to pinpoint important events. The graphs of past results come from the program's database.

## The Database

The database is a compilation of selected variables from past experiments. As each experiment ends, the student is asked if he/she wants to include results in the database. The tank database will show the student graphs of cost per kilogram of weight gain (*y* axis) versus temperature, oxgyen, or percent protein in the feed (*x* axis). The pond experiments will graph average fish weight, total harvest or profit (*y* axis) against stocking density (*x* axis). The database can be cleared, or individual data points can be deleted from it. Again, in the Apple II program, these graphs are presented in low-resolution graphics.

## The Unknown Fish

*Fish Farm* contains a fairly accurate model of single-crop catfish culture as it was practiced in the southern U.S. in the early 1980s. The *Fish Farm* disk also contains 26 "unknown" fish, called Fish A through Fish Z. These fish are fictitious but based on real species, and the knowledgeable user will be able to spot unknown fish with strong resemblances to catfish, carp, rainbow trout, and *Tilapia*. Carp- or catfishlike fish have moderate requirements; a troutlike fish requires clean, cold water and protein-rich feed; a *Tilapia*-like fish prefers warm water, is pollution-tolerant, and must be cultured at high stocking densities. The unknown fish may be loosely classified as follows:

**Carp Group:** C, I, J, M, Q, S, W

**Trout Group:** B, D, H, L, O, T, X, Z

***Tilapia* Group:** A, F, G, K, P, V, Y

**Other Types:** E, N, R, U

The unknown fish have a wide variety of optimum culturing conditions. Stressful oxygen concentration ranges from 1.8 to 6.7 mg/l; preferred feed protein content ranges from 8% to 63%; optimum stocking density ranges from 200 fish/hectare to 5.5 million fish/hectare. But despite their diversity, all the fish will yield profits of about $250,000 per growing season if they are cultured correctly. Correct culturing parameters of each of the unknown fish are shown on page I-7.

When the student requests each of the unknowns, the program will display a narrative description which will give culturing hints. For example, the description of Fish T (from the trout group) reads:

*Fish T is found in clean streams on Vancouver Island on the Pacific coast of Canada. It is a slow-growing, cold-water, disease-prone fish with rigid requirements for clean water and high oxygen. Fish T is expensive to raise (its fingerlings cost $1.61 each, and it accepts only the most high-protein feed). But its meat is a luxury product and will richly reward the culturist who can meet its high water quality requirements.*

## Optimal Culturing Conditons of *Fish Farm's* "Unknown" Fish

| Fish | Mar. Wt. | Price | Grdwtr[1] | Oxygen | Protein | Harvest[2] | Fish/Hectare[3] | Comments |
|---|---|---|---|---|---|---|---|---|
| Catfish | 454 g | $1.65 | no | 5 mg/l | 30% | ES | 7,500 | |
| Fish A | 320 g | $0.99 | no | 1.8 mg/l | 16% | MW | 30,000 | MW requirement only apparent at stkg dnsty near optimum. |
| Fish B | 150 g | $5.58 | yes | 5.9 mg/l | 59% | ES | 700,000 | |
| Fish C | 900 g | $2.34 | no | 4.6 mg/l | 24% | ES | 3,750 | |
| Fish D | 454 g | $7.20 | yes | 6.0 mg/l | 41% | ES | 275,000 | |
| Fish E | 1000 g | $8.90 | no | 4.5 mg/l | 28% | ES | 200 | Low stocking density |
| Fish F | 227 g | $1.12 | no | 2.8 mg/l | 24% | ES | 42,000 | Variable profits |
| Fish G | 120 g | $2.69 | no | 3.5 mg/l | 45% | ES | 35,000 | |
| Fish H | 500 g | $2.99 | yes | 6.5 mg/l | 48% | ES | 500,000 | |
| Fish I | 700 g | $1.32 | no | 3.0 mg/l | 22% | ES | 12,500 | |
| Fish J | 600 g | $1.26 | no | 3.0 mg/l | 21% | ES | 18,000 | Hot weather = stress, disease |
| Fish K | 150 g | $2.52 | no | 2.8 mg/l | 38% | ES | 28,000 | Variable profits |
| Fish L | 360 g | $4.50 | yes | 3.8 mg/l | 42% | ES | 325,000 | |
| Fish M | 1000 g | $1.55 | no | 3.9 mg/l | 8% | ES | 4,500 | |
| Fish N | 400 g | $4.11 | no | 4.0 mg/l | 32% | ES | 4,000 | Variable profits |
| Fish O | 454 g | $5.60 | yes | 5.8 mg/l | 52% | ES | 400,000 | |
| Fish P | 225 g | $2.19 | no | 5.2 mg/l | 34% | ES | 13,500 | Variable growth rates |
| Fish Q | 700 g | $1.39 | no | 4.5 mg/l | 18% | ES | 14,000 | Growth stops in hot weather |
| Fish R | 150 g | $1.42 | yes | 4.1 mg/l | 39% | ES | 5,500,000 | Best profit around $210,000 |
| Fish S | 600 g | $1.19 | no | 3.6 mg/l | 18% | ES | 15,000 | |
| Fish T | 454 g | $4.61 | yes | 5.5 mg/l | 55% | ES | 400,000 | |
| Fish U | 150 g | $14.49 | no | 2.7 mg/l | 63% | ES | 1,250 | Variable profits |
| Fish V | 350 g | $2.09 | no | 3.8 mg/l | 15% | ES | 18,000 | Variable profits |
| Fish W | 900 g | $2.54 | no | 4.3 mg/l | 20% | ES | 2,800 | Best profit around $330,000 |
| Fish X | 454 g | $7.55 | yes | 6.7 mg/l | 55% | ES | 200,000 | |
| Fish Y | 200 g | $2.53 | no | 4.5 mg/l | 41% | MW | 20,000 | |
| Fish Z | 1362 g | $1.51 | yes | 3.4 mg/l | 26% | ES | 350,000 | Variable at high stkg densities |

[1] Yes = use 190% groundwater input per day; no = use no groundwater. Groundwater inputs less than 190% are ineffective.

[2] ES = harvest at end of season (November 5); MW = harvest when fish reaches marketable weight

[3] Stocking density which will produce best profit, around $250,000/season, when other variables are correct.

At Clemson, three unknown fish are used per semester (which will allow the unknown fish to be used for nine semesters before they begin to repeat). If those fish are Fish A, B and C, then every lab section has two teams working on Fish A, two teams on Fish B and two teams on Fish C. Consultation between two Fish A teams (for example) is encouraged in the same way that consultation between different investigators working on the same organism is a normal part of science.

Having several teams work on one fish also allows competition between those teams for the highest profits on the production run. At Clemson, all the production run records are collected (blank form in Appendix B) and results and culturing conditions for the most profitable teams are posted. This not only gives some public recognition to the teams with the highest profits, but also allows all teams using the same fish to compare their results with those of the "winners" when they write their reports. This profit competition should not be the major emphasis of *Fish Farm* (the major emphasis should be on sound scientific procedure and well-done reports), but friendly profit competition does heighten interest and motivation.

Another way to use *Fish Farm* in class is to divide responsibility for the tank experiments so that one set of teams does the temperature experiment, another set does the oxygen experiment and a third set does the feeding experiment. This arrangement can present a good opportunity for students to give oral reports because the temperature teams have to be informed about the oxygen and protein results, the oxygen teams have to know the temperature and protein results, and the protein teams have to know the temperature and oxygen results.

## Student Success with *Fish Farm*

One way of gauging the ability of students to solve the *Fish Farm* problem is by looking at their profits. Recall that if a fish is cultured under optimal conditions, it should yield a profit of about $250,000 per growing season. Experience at Clemson shows that if students make a reasonable effort and follow directions, most of them will reach profits of this magnitude. In fall of 1992, all 26 unknown fish were tested with 456 Clemson students in 24 sections. The exercise was supervised by 13 lab instructors, 9 of whom were new TA's who had never used *Fish Farm* before. The frequency distribution on page I-9 shows the profit per growing season during the final production runs. Results from all 26 unknown fish have been pooled.

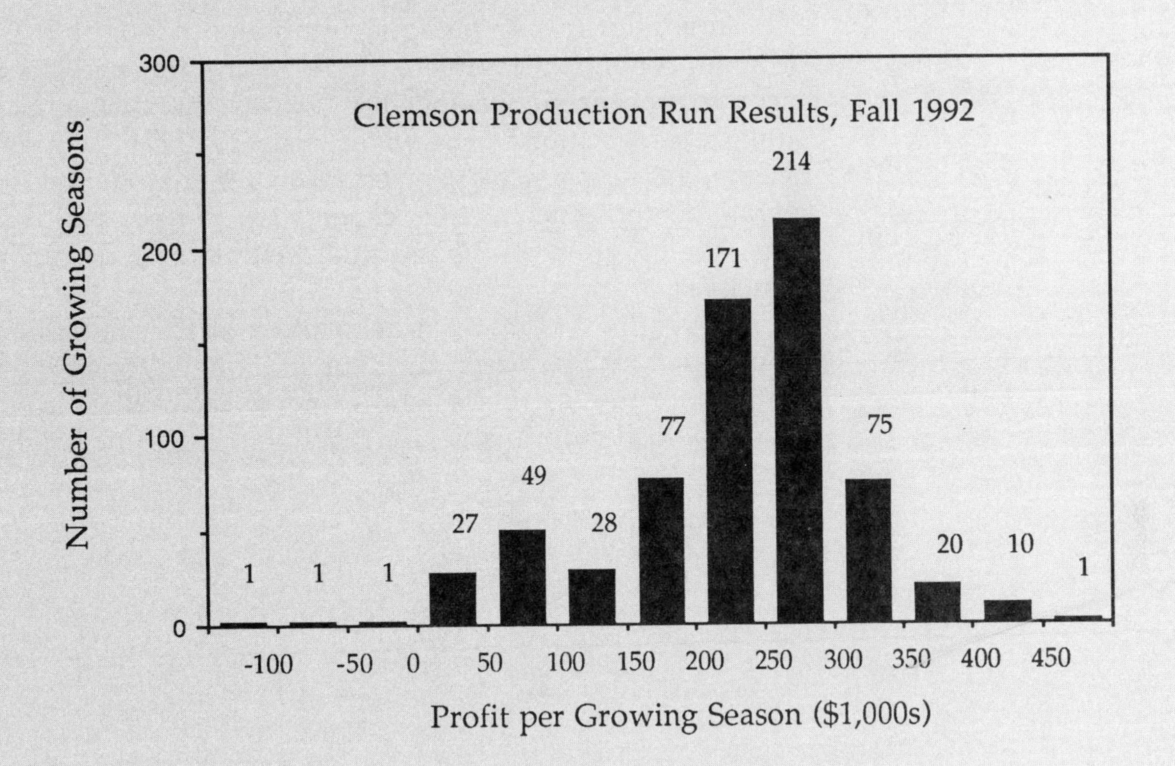


Note that, out of 675 student-simulated growing seasons, only 3 seasons failed to make a profit, and only 79 seasons (12%) made less than $100,000 per growing season. These results did not occur because the *Fish Farm* problem was easy. At the beginning of their work with the program, these same students were making no profits at all (and were becoming discouraged). But they were ultimately successful because they consulted with their instructors and with each other, and had the patience to systematically determine all the correct culturing conditions.

The figures above are from a supervised classroom situation. What about students using the program on their own? Anecdotal evidence suggests that unsupervised students will be less successful, mostly because they tend to go right to the computer keyboard without reading the background material, (or, in some cases, even the directions). Many of these students try to guess the correct culturing conditions, can't make money, and then report that the problem was "frustrating." This is why it is recommended that *Fish Farm* be used in class, or failing that, outside of class with assistance readily available.

# Using *Fish Farm* as a Population Ecology Exercise

*Fish Farm* is primarily an aquaculture simulation, but the program contains a complete model of a pond ecosystem, including stochastic weather for Charleston, South Carolina; sunlight penetration in the water; organic decay; nitrogen cycling; and dynamics of bacteria, phytoplankton, and invertebrate benthos such as oligochaetes and chironomid larvae. Thus, the program can be used to illustrate many ecological principles.

Appendix C contains a description of the simulation model used by *Fish Farm*, and Appendix D has some suggestions for the instructor who wants to move beyond profitable fish culture and use *Fish Farm* as a pure ecology exercise.

INSTRUCTOR CHAPTER 2

# The First *Fish Farm* Lab

The objectives of the first *Fish Farm* lab are to:

1. Give an introduction to *Fish Farm* and to aquaculture (30 minutes);

2. Do a familiarization exercise with the program, using catfish (30 minutes);

3. Introduce the idea of independent, dependent and controlled variables (15 minutes);

4. Introduce and complete the *Fish Farm* temperature, oxygen and feeding exercises (75 minutes).

**Before Class**

Students should read chapters 1 and 2 in the student manual. Worksheet 1 (page 20) will be done in class.

## Introductory Material

1. Explain that one of the purposes of a science course is to train the students how to use scientific problem-solving skills. The purpose of *Fish Farm* is to allow the students to work on some basic skills such as making graphs and recording data, and also to give them practice at solving a complex, multifactor problem: making aquaculture profitable. Review the schedule of activities on page 5 of the student manual to give the students an overview of what's coming.

2. Give the students the *Fish Farm* scenario (that they are the director of research for an aquaculture company, that they must determine economically optimum culturing conditions for a new fish, etc., as in the manual). State that the objective is to have the greatest possible number of fish gaining weight as rapidly as possible.

3. Explain that this is a complicated problem because many factors must be correct at the same time. For example, if the food, the stocking density, etc. are all correct but the fish is at the wrong temperature, it will not be able to grow and the enterprise will not make a profit. This will bring you to the orientation exercise.

## The Orientation Exercise

In this exercise, challenge the students to raise catfish profitably without the systematic experiments in the rest of the *Fish Farm* exercises. Hopefully, most of them will **not** make a profit, and hopefully this will convince them that the problem-solving procedures in the rest of the *Fish Farm* exercises will be useful to learn. Also, the orientation exercise helps to allay the fears of the computerphobic and give the students a little familiarity with the program

The actual exercises are described on the "*Fish Farm* Orientation" pages of the student manual (pages 6-8).

4. Get each group seated at a computer, and make sure they all have disks (if you don't have *Fish Farm* loaded onto hard drives). Start the program according to the directions in the "Installing *Fish Farm*" section in Appendix A of this *Instructor's Guide.*

5. Tell them to start following the directions on page 6 of the student manual. Explain the directions to them, but try to resist the temptation to whisper hot tips on protein percentages and the like.

6. Expect a lot of questions and a moderate amount of chaos, but after 20-30 minutes, call a halt and ask for profits. The best they can expect to do is $300,000 or a little more, so don't be fooled by claims of millions and billions. But if a group **does** hit it just right, ask them how they did it. If it was a series of lucky guesses, use the fact that other teams were probably not so successful to show that lucky guesses do not occur routinely. If they had virtuously read their manual and had found the correct information, commend them for their scholarship, triumphantly point out that knowledge is the key to successful experimentation, but warn that when

they get their unknown fish, there will be no information in the manual.

7. Conclude that they will be taught a problem-solving method which will work with almost any complex problem like this, and which they can use for the rest of their lives: controlled experimentation (hold all factors but one constant, etc.).

   To demonstrate that it *is* possible to make a profit with catfish, tell them to try the following conditions:

   Stocking density of 7,500 fish/hectare
   30% protein feed, fed at a varying rate
   No input of groundwater (0%/day)
   Aerator goes on when oxygen goes below 5 mg/l.

These conditions should give a profit of about $250,000 per growing season. Once all the groups have run these conditions, ask each of them to tell the class what their profits were. The profits of each group will be different. Use this fact to make the point that different *Fish Farm* simulations (even using the same conditions) produce different results. This may bewilder some students at first; they think of a computer program as a big calculator which should give the right answer if they push the right keys. Explain that one of the major tasks of a scientist is dealing competently with variable results, as they will find out as they start the experiments on their "unknown" fish.

## Variables

The tank experiments are partially an experimental design exercise, and if the students are going to benefit from them, they will need some background on the different types of variables.

9. Write "independent variable," "dependent variable," and "controlled variable" on an overhead or a blackboard, but leave plenty of space for writing under each one.

10. Define each type of variable (pages 35-36 of the student manual) and ask for an example of each type of variable

from the orientation experiment. You'll probably get stocking density as the independent variable, profit as the dependent variable, and a variety of answers for the controlled variables.

**11.** Emphasize the importance of having only one independent variable at a time. They will need to be aware of this as they perform the tank experiments.

## The Tank Experiments

The conclusions which the students draw during the tank experiments will be used throughout *Fish Farm,* so tell them to be careful.

**12.** Assign unknown fish to the different groups. At Clemson, three fish per section (with two teams per fish, each team doing all experiments) and one fish per section (with a single team only doing one of the three tank experiments) have both been used.

**13.** If the tank experiments are going to be split between the teams, explain that the first series of experiments will be a cooperative section effort. Assign equal numbers of teams to do temperature, oxygen, and protein percentage in the feed.

**14.** Use the transparencies of Figures 1, 2, and 3 (masters in Appendix B) to summarize how fish in an aquarium respond to changing temperature, oxygen, and protein.

**15. Have the students turn to pages 18-19 of the student manual (transparency also included in Appendix B). Strongly emphasize the objectives of the tank experiments and explain why the results will be important to their culturing efforts.**

**16.** To test their understanding, have the students turn to Worksheet 1 on page 20 of the student manual and ask for answers to the questions. They can consult the manual for this assignment. Give them a few minutes, and then go over the results. A discussion guide follows.

## Worksheet 1 Discussion Guide

1. Raise a fish with a high groundwater input only if it cannot tolerate temperatures of 25° to 30°. These temperatures will be common in static (without groundwater) ponds in coastal South Carolina in the summer.

2. **Don't** use groundwater if it is avoidable. Table 2 on page 12 of the student manual shows that even relatively modest groundwater inputs will sharply limit the number of hectares of ponds that will be useable. There is only a limited water supply for the ponds; the more water any one pond gets, the fewer ponds the well can support.

3. Aeration is expensive, so turn on the aerators at the **lowest** oxygen which still allows the fish to grow adequately.

4. FCR stands for "feed conversion ratio." This ratio is feed given/weight gained, and so it is better if it is **low** (little feed given, much weight gained). In catfish culture 2.0 is the standard to shoot for, and when catfish are young and conditions are optimal, FCR's of 1.0 have been reported. But keep in mind that an FCR of 1.0 is not ecological magic-the weight gain is 80% water, and the feed given is only 10% water. Therefore an FCR of 1.0 corresponds to a feed-to-biomass conversion rate of 22%, still far better than the 1%-3% seen in feedlot cattle.

5. Don't overfeed protein because it raises feed costs and it pollutes the water with ammonia. Ammonia is toxic, and its oxidation to nitrate consumes almost four times its weight in oxygen. *Fish Farm* does contain this oxygen-gobbling nitrification reaction. Both low oxygen and high ammonia can put fish off their feed, and even kill them, by direct toxicity and by triggering disease outbreaks.

17. If the tank experiments are going to be split up between the teams, tell the temperature team to turn to page 21,

the oxygen team to turn to page 26, and the protein teams to turn to page 28 of the student manual. If all teams are going to do all experiments, everyone should turn to page 21.

**18.** As they begin, remind the students that their data sheets will be turned in with their final reports and will be 30% of their final report grade. Page 87 of the student manual summarizes the characteristics of a good data sheet and the points attached to each characteristic.

**19.** **Be sure that teams doing temperature and oxygen use a *constant* feeding rate and that the protein teams use a *varying* feeding rate.**

**20.** As the teams try one temperature, oxygen, or protein after another, be sure that they're adding their results to the program's "database." Have teams look frequently at the graphs from the tank database, because these will suggest additional values they have to try. Don't forget to advise replication.

**21.** Students may be bewildered by the fact that duplicate experiments turn out differently from one another. Explain that this occurs because the program gives each new batch of fish a slightly different growth rate and a slightly different resistance to stress.

**22.** *Fish Farm* scales its database graphs so that the range of x values and the range of y values take up the whole area of the graph. This may cause a problem. If teams generate a graph which looks like a random scatter of points, look at the range of values on the y-axis and the x-axis. If these ranges are small, all teams are seeing is random fluctuation around a constant value (for example, fluctuation around an optimum response to oxygen when none of the oxygen values tried so far stresses the fish). When teams try more extreme values of temperature, oxygen or protein, the true shape of the curve will rapidly emerge because the new extreme y values will flatten the former variability into a tiny region of the graph.

**23.** If teams make a mistake and then correct it, advise them to clear the tank database before adding the subsequent results. Clearing the database is an option on the end-of-experiment menu.

**24.** When a team finishes with one tank experiment (e.g., temperature) and goes on to the next, the tank database must be cleared. If this isn't done, there will be a confusing "spike" of points on their graph as they begin the next experiment. To understand why this occurs, imagine that a team does a temperature experiment and collects 8 cost per kg of weight gain values at various temperatures, all using 10 mg/l oxygen. The team then goes on to the oxygen experiment without clearing the database. As they start, 8 cost points left over from the temperature experiment will be plotted on their oxygen graph, all piled up at 10 mg/l. There will be no indication on the oxygen graph that these were all taken at different temperatures.

**Homework**

For homework the students must read Chapters 3 and 4 in the *Fish Farm* student manual (pages 35-64) and complete Worksheets 2 (page 49) and 3 (page 65). These worksheets must be turned in at the *beginning* of the next week's lab.

CONTENTS

FISH FARM

# Lab 1

CHAPTER 1

# Introduction to *Fish Farm*

Learning the methods of scientific investigation is an important part of a biology education. A biology student should learn how to record and present data, should be able to determine when a data set is adequate for answering a question, and should be familiar with scientific methods of presentation to colleagues. We will attempt to teach you these skills by asking you to perform a series of exercises with *Fish Farm*, a computer simulation of the biology and economics of a commercial aquaculture enterprise.

Using a computer simulation, you can do experiments without any concern for how long they might take, the equipment or skills they might require, and their cost. For this reason, computer simulations are a common tool of science. Space voyages are preceded by years of simulation studies, and the data recorded on the flight are used to refine the computer models to improve their predictive power. Large-scale environmental problems such as ozone depletion, acid rain, and global warming are intensively researched using simulations. Simulations are the *only* practical way to study problems like the possibility of "nuclear winter" and the theory that the dinosaurs became extinct due to the impact of a giant meteor on the earth. *Fish Farm* will give you insight into how simulations can quickly explore a problem and find the probable solution to it.

We think you will find that *Fish Farm* is an interesting and beneficial part of your introductory biology course.

# Overview of *Fish Farm*

In *Fish Farm*, you must determine the correct culturing conditions for a fish, and then raise this fish as profitably as possible.

You will play the role of the head of the research division of an aquaculture company. Until now, your company has been a profitable channel catfish "grow-out" operation. That is, it purchases catfish fingerlings (about 15 cm long, weighing about 25 g each), stocks them in ponds in late March, and feeds them until early November, when they are about 30 cm long and 600 g each. Then it processes them into fillets and sells them to restaurant chains.

Recently the company has acquired a new variety of fish which appears to have commercial promise. The company plans to raise this fish in a production unit near Charleston, South Carolina; the unit has 200 hectares (about 500 acres) of ponds. But before this large facility is commited to the project, you must determine how to culture the fish in order to maximize profits.

Your problem actually contains several sub-problems:

First, you must **determine the physical tolerances of the fish with respect to temperature and dissolved oxygen**. If the fish can tolerate wide fluctuations in these variables and do not require very clean water, they can be grown in ponds with little water flow through them. On the other hand, if the fish require a narrower range of temperature and oxygen, they will have to be grown in ponds or raceways with rapid water flow. But your water supply is limited, and the requirement for a rapid flowthrough will cut down on your total production and profits.

Next you will **investigate diet**. The fish will be fed pellets of artificial diet, and the cost of the feed will probably be your biggest single cost. The most expensive component of the diet is protein. You must determine the lowest percent of protein that assures fish health and rapid weight gain.

The next step is to **determine stocking density**, or the number of fish to put in each pond. An accurate decision here is vital. Your objective will be to stock the fish at the highest possible density that will allow a successful harvest. You must also determine when to harvest the fish to maximize profit.

Lastly, you must **test your recommended culturing conditions with a "production run"** in which you will simulate five years of commercial operation.

While a profitable production run will be a good indicator of careful, orderly experimentation, your grade will be based on a *Fish Farm* report which will summarize your culturing recommendations and the data which support them.

You will spend two laboratory periods with *Fish Farm;* each one will be preceded and followed by a homework assignment:

**Before Lab 1:** Read Chapters 1 and 2 of the *Fish Farm* manual.

**Lab 1:** Orientation exercise and "tank" experiments. Students will briefly explore the operation of the program and test their ability to raise catfish profitably. Worksheet 1 (page 20) will be completed in class. Then an "unknown" fish will be assigned, and each team of 3 or 4 students will determine this fish's response to temperature, oxygen, and feed protein.

**Homework:** Read Chapters 3 and 4 of the *Fish Farm* manual. Then *each student* will do Worksheet 2 (page 49) and Worksheet 3 (page 65). Worksheet 2 will use the *Fish Farm* data collected during Lab 1.

**Lab 2:** Each team will determine harvest time and optimum stocking rate for their fish. Using this information, each team will perform the production run.

**Homework:** *Each student* will write a final report (directions in Chapter 5).

The next few pages will allow you to experiment with *Fish Farm* so you can see how the program operates.

## The *Fish Farm* Orientation Exercise

In this "get acquainted" session, you will explore the response of *catfish* (not one of the unknown fish) to whatever culturing regimes you decide to test. You will also find out how difficult it can be to make a profit without basing your culturing conditions on a) knowledge of the system and b) an orderly program of experimentation.

*Fish Farm* simulates a farm with 200 hectares (495 acres)

of ponds. A superior catfish farm of this size should be able to make a profit of $500,000 per year. But real catfish farms usually use a complex "multiple harvest" system which increases their profitability. For simplicity, *Fish Farm* uses a "single harvest" system, and in this system profits should exceed $200,000 per year and may occasionally exceed $300,000 per year. So, as you perform the following exercise, remember that your objective is to manipulate culturing conditions to bring your profits to about **$250,000 per year**.

What will make the exercise difficult is that there are so many culturing variables (how many fish to put in the pond, what to feed them, how much to feed them, etc.), and **all** of the variables must be correct at the same time. If one variable is wrong (for example, everything is correct except the type of feed), you will fall far short of your goal.

## Orientation Exercise Procedure

**Note:** The details of *Fish Farm* are different on Apple II, IBM, and Macintosh computers. When in doubt, follow the directions on the screen.

1. Place your *Fish Farm* disk in the disk drive of the computer and turn the monitor and the computer on.

2. After a title screen, you may be asked to adjust the program speed so a second timer ticks at the correct rate. Follow the directions on the screen.

3. If you are working with an Apple IIe or Apple IIGS computer, next indicate that you wish to work with **outdoor ponds**. This question will come later for IBM and Macintosh users.

4. Select **Catfish** for the experiment. Read the background information presented, and then accept catfish as your fish.

5. If you are working with an IBM or a Macintosh computer, indicate that you wish to work with **outdoor ponds** after your choice of fish.

6. This will bring you to the "Pond Menu." This menu lists the culturing decisions you will have to make:

   a. how many fish to put in the pond
   b. the protein content of the feed
   c. how much feed to give
   d. whether to use feed medicated with antibiotics
   e. whether to use a flow of groundwater through the pond
   f. when to mechanically aerate the ponds
   g. when to harvest the fish

7. Select **stocking density** and put in what you regard as a high but not overcrowded number of fish per hectare (a hectare is an area 100 m by 100 m--about 2.5 acres). Try between 1,000 and 500,000 fish per hectare. The most profitable stocking density will depend on your choices below.

8. Select **feeding instructions**. Here you must first decide how much protein to put in the feed. Feeds ranging from 10% to 80% protein are used in aquaculture.

9. Next, the program will ask you whether you prefer to feed at a **constant** or a **varying** rate. Feed at a **varying** rate.

10. Do you wish to add antibiotic to the feed? This can stop disease outbreaks, but it makes the feed more expensive. If you don't wish to use antibiotics, ignore this option.

11. Next, do you wish to flush the pond with groundwater each day? While most fish are raised in static ponds (with no water flowthrough--0% input/day), some fish require high water input in order to grow. If you don't wish to use groundwater, ignore this option. If you do wish to use groundwater, you may put in 0%-190% of the volume of the pond per day.

12. Do you wish to aerate the water with paddle wheels when the oxygen gets low? If you don't wish to use aeration, ignore this option. If you wish to use aeration, the program will ask you at what dissolved oxygen concentration you wish the aerators to be turned on. Use a dissolved oxygen concentration lower than 8 mg/liter as the "trigger" concentration.

**13.** When should you harvest the fish? For catfish, the correct option is "End of Season," which is the option specified when the program begins. So leave this one alone.

**14.** Press **ESC** and the simulation will begin. There are two different types of screens you can observe:

a. The digital screen shows pond conditions, fish weight gain, and profit. Profit will not be positive until the value of the fish exceeds expenditures.

b. The graphics screens show graphs of water temperature (press "T"), dissolved oxygen concentration (press "O"), mean fish weight (press "W") and profits (press "P"). The last graph will probably be the best curve to watch.

But the program requires no action on your part to run properly.

**15.** On November 5, the program will stop automatically and display the results of the experiment. The last line shows your profit (or loss). Record these profit/loss figures for later discussion. If any fish died, an additional screen will give you a "Pathologist's Report" which shows the causes of death.

**16.** Indicate that you wish to **discard** the results, that you wish to **continue to the main menu**, and then that you wish to perform **another experiment**. Indicate that you desire to use **outdoor ponds** again. Then repeat steps 7-15 with whatever additional conditions you wish to try.

When you have tried the last set of conditions you wish to test, discuss the results with your instructor. Your profits will probably not come close to the $250,000 goal, but this is not surprising. The aquaculture problem is very complex because so many conditions must be correct at the same time, and the conditions which are used in commerical aquaculture were derived from decades of experience and research. One of the goals of biology is to teach you the scientific approach--a step-by-step procedure for dealing with complex problems.

a. Decide on the **dependent variable** (such as profit) which you want to study, and decide which **independent variables** (such as stocking density, food amount, and so forth) influence it.

b. Perform an experiment where you hold all of the independent variables except one constant (e.g., hold food, aeration, groundwater, and all other variables the same while you systematically vary stocking density). Observe the effect on the dependent variable (profit).

c. When you have determined the best value for the independent variable, leave it at that value, choose another independent variable, and perform the same steps with it (e.g., if you found that the most profitable stocking density was 30,000 fish per hectare, leave stocking density at 30,000 and start systematically varying food protein percentage).

d. Continue the same procedure until you have found the most profitable values for each independent variable. At that point, you probably have determined the most profitable set of culturing conditions.

Most people do not use this logical "scientific" procedure in daily life because they think they can use their knowledge of the system as a shortcut to guess the optimum values. While ignorant guessing rarely produces good results in complex problems, knowlege of the system gives either the scientific or non-scientific investigator important advantages. One of the worst things a scientist can say about another scientist's experiment is that it is "naive"--that it shows ignorance of well-known facts about the investigated system. Because knowledge is so important, this manual contains both specific background material that will allow you to understand each of the exercises, plus a more general overview of aquaculture in the Appendix at the end of the manual.

Therefore, during the rest of the *Fish Farm* exercises, we will give you the opportunity to learn two things:

a. the scientific method of problem solving and

b. sufficient knowledge of aquaculture to make the scientific method an efficient solution for the *Fish Farm* problem.

CHAPTER 2

# The Tank Experiments

In this series of experiments, you will use controlled conditions in indoor tanks to determine the temperature tolerances of your fish, the oxygen concentration at which it first begins to grow poorly, and the feed protein content which produces the cheapest weight gain.

## Temperature Experiment Background

Temperature affects every chemical reaction in an animal. Since fish are usually at almost the same temperature as the water surrounding them, it is clear that environmental temperature will have a large role in fish health and growth.

Extremely hot or cold temperatures or sudden temperature changes cause a serious condition called thermal shock. This can kill fish by causing changes in tissue chemistry and degeneration of red blood cells, and by depressing the fish's immune system and allowing an outbreak of disease.

Less extreme temperature conditions can also influence fish growth. Growth usually increases with temperature until the optimum temperature is reached, and then declines rapidly as thermal stress begins to occur. For example, carp in outdoor ponds in chilly Germany (mean annual temperature only 8° C!) require almost 3 years to reach 700 g. In outdoor ponds in semitropical Israel, growers can harvest *two* crops of 700 g carp *per year*.

Different fish have different preferred temperature ranges. Rainbow trout grow best at 10°-18° C, stop feeding by 22° C, and by 25° C are dying from heat stress. Channel catfish start to grow well when the water is warmer than 21° C, have an optimum temperature around 28°-30° C, and can tolerate up to 36° C before they start to die. And Nile *Tilapia* can tolerate up to 42° C (107° F!).

## Controlling Temperature with Groundwater Input

Large outdoor ponds basically follow the temperature regime of their environment, and cannot be heated or cooled without extravagant expense. The simulated *Fish Farm* outdoor ponds are located near Charleston, South Carolina, and in Charleston's subtropical, maritime climate, pond temperatures vary from about 6-10° C in winter to 28-35° C in summer. Some fish may not be able to stand these temperatures, but a limited kind of temperature control can be achieved by pumping constant-temperature groundwater through the ponds. The groundwater is at 18° C, and the greater the influx you provide, the closer you will hold the pond temperature to 18° C. The temperature-moderating effect of groundwater influx is shown in Table 1.

**Table 1.** Temperature of outdoor ponds between March 21 and November 5 as a function of the amount of groundwater input. Input is expressed as a percentage of the pond volume which is pumped in per day.

| Inflow (%/day) | Lowest Temp. | Highest Temp. | Largest 12 Hr Fluc. |
|---|---|---|---|
| 0% | 10.0° C | 32.7° C | 12.0° C |
| 50% | 12.1° C | 27.7° C | 8.3° C |
| 100% | 14.0° C | 24.5° C | 5.5° C |
| 150% | 16.5° C | 21.2° C | 3.0° C |

Note the line corresponding to 0% input. This shows the normal temperatures you can expect in a "static" (no water input) pond.

Use of groundwater is not free. The production unit well has a limited output of 220 liters/sec, and so groundwater input will reduce the number of hectares of ponds you may use in the production run.

**Table 2.** Effect of groundwater input on the hectares of production ponds which can be supplied by the production unit well.

| Input (%/day) | Hectares Supplied |
|---|---|
| 0% | 200.0 |
| 1% | 190.1 |
| 5% | 38.0 |
| 50% | 3.8 |
| 100% | 1.9 |
| 190% | 1.0 |

Small groundwater inputs (e.g., less than 100%/day) will probably not have much influence on water temperature or waste removal. Groundwater is used either at highest possible rate (190%/day in *Fish Farm*) or not at all. Therefore, be aware that depending on large groundwater inputs will sharply limit the area of ponds the well will support. You will have to compensate for the smaller pond area by using higher stocking densities. But you **must** use groundwater inputs if your fish cannot tolerate the normal temperatures in static ponds.

## An Example of a Temperature Experiment

You will determine the temperature tolerances of your fish in indoor tanks with rapid water flowthrough and carefully controlled conditions. The tank experiments will last from March 21 to April 30.

In the experiment in Figure 1, 10 catfish fingerlings (27 g each) were fed 5% of their body weight per day (more than they can eat). Figure 1 graphs the mean weight and the cost of feed per kg of weight gained against temperature. The results for your fish should show a similar pattern, although the critical temperatures might be different.

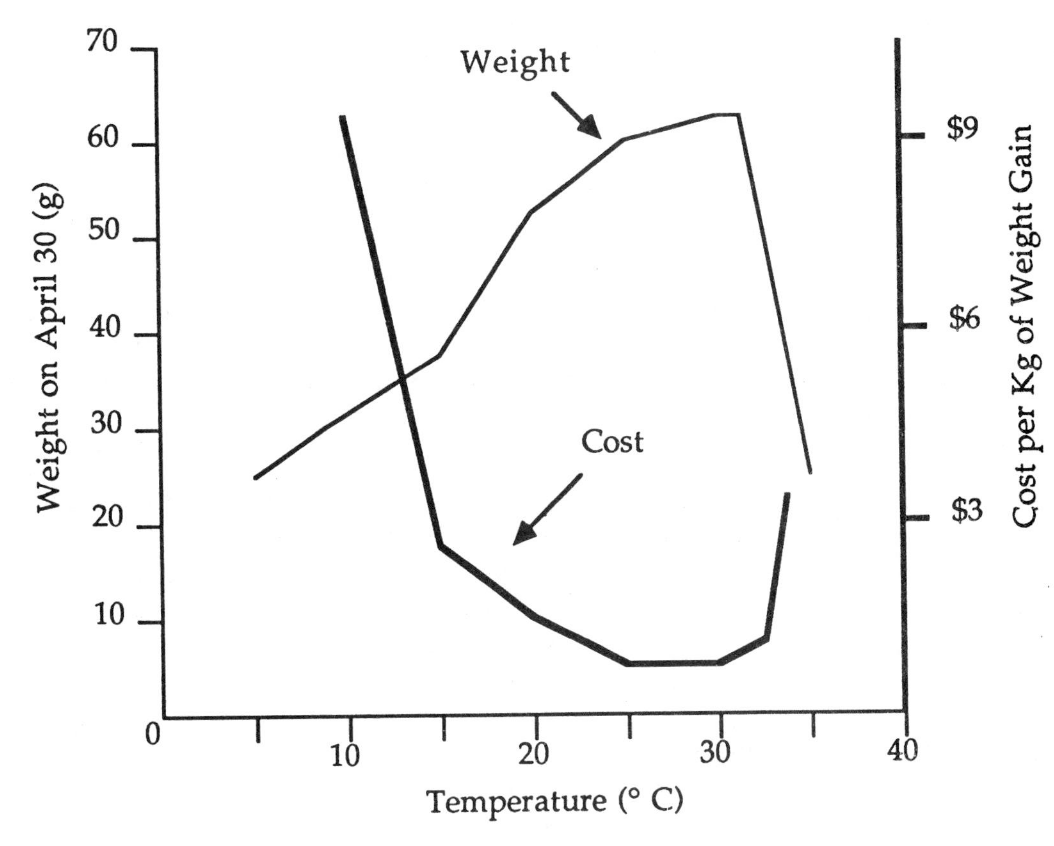


**Figure 1**. Growth of catfish as a function of temperature. Oxygen was set at 10 mg/l, and feeding at 5% of biomass/day with 50% protein feed.

The increasing cost at temperatures away from the optimum (25°-30° C for these catfish) is caused by the fact that the fish eat less at those temperatures, and so a larger percentage of the food given is wasted. However, it is clear from the results that catfish can tolerate pond temperatures of 25°-30°, so the conclusion from this temperature experiment would be that groundwater input is **not** required for raising catfish.

## Oxygen Experiment Background

Oxygen is frequently the most limiting (and vital) resource in aquaculture. Air is 21% oxygen, and terrestrial organisms almost never encounter oxygen-depleted air. But oxygen gas is rarely more than 10 mg/l (0.001%) in water and is commonly 6-8 mg/l in warm surface water. If water is polluted or shut off from exchange with the air, its dissolved oxygen can entirely disappear. Fish have adapted to these conditions in nature and will leave water with low oxygen if they can. But your fish are trapped in their pond, and so you must ensure that the oxygen there does not get dangerously low.

The oxygen in aquaculture ponds is mainly consumed not by the fish themselves but by bacteria and algae in the water (one study found that only 9% of the oxygen consumption in a catfish pond was due to the fish). The bacterial load of the water builds up mainly because feeding the fish is introducing large amounts of organic matter, and uneaten food and feces decay rapidly. These problems tend to get worse as the season progresses because the fish are growing. In March, a 1 hectare pond with 10,000 fish/hectare has 270 kg of fish and requires about 8 kg of feed each day. But in September, the pond has 5,000 kg of fish and might require 100 kg of feed/day. This tremendous influx of organic matter is also occurring when pond temperatures are high and the ability of the water to hold oxygen in low. The stage is set for an oxygen depletion crisis in late summer and early autumn, just before harvest.

Low oxygen stresses fish, causes them to stop feeding, and may cause disease. If oxygen gets very low, the fish can suffocate. Mechanical aeration of the water (usually with a paddle wheel device) can avoid catastrophic mortality, but it is expensive and should be used as infrequently as possible. In your *Fish Farm* pond experiments, you may set an aerator to turn on whenever dissolved oxygen reaches the concentration which begins to stress the fish, but this concentration is unknown. Determining the concentration is the main purpose of the oxygen experiment.

How much oxygen is enough? Catfish *begin* to be stressed as the dissolved oxygen reaches 5 mg/l, and they die within a few hours when it reaches 1 mg/l. If you were growing catfish with mechanical aeration, you would want to start your aerators when the dissolved oxygen reached 5 mg/l. You would not want to set the aerators to go on at a lower value because of the danger of fish death, and aerating when oxygen was above 5 mg/l would be expensive and not result in additional fish growth.

Using a tank experiment similar to the one for temperature, you will determine the reaction of fish to chronic low oxygen conditions, maintained from March 21 to April 30. Your results should resemble the results shown in Figure 2 for catfish, although the fish you are assigned might have a different tolerance for low oxygen.

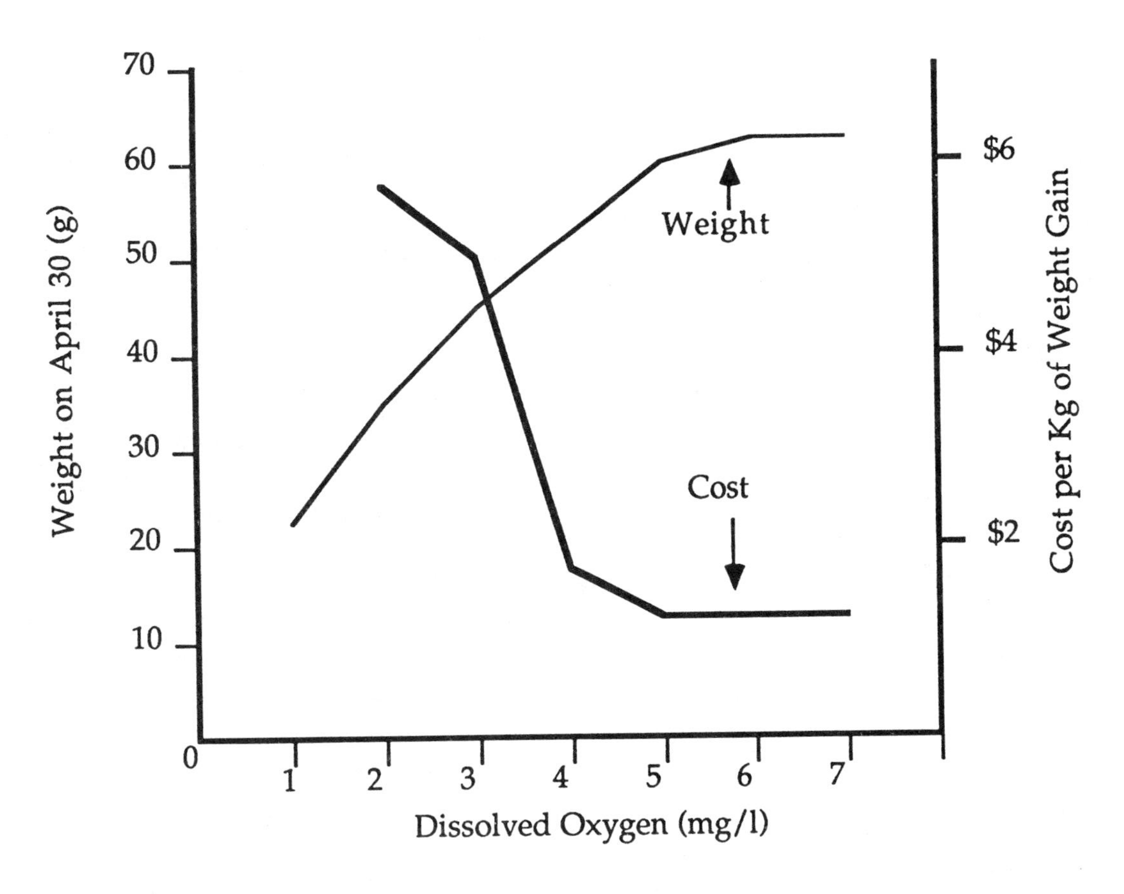


**Figure 2.** Growth of catfish as a function of dissolved oxygen. Temperature was set at 25° C, and fish were fed 5% of biomass/day with 50% protein feed.

## Feeding Experiment Background

Careful attention to feeding is important for aquacultural success. On the one hand, obviously your fish must get a sufficient amount of the kind of food they need before they can grow. But on the other hand, feed will be the most expensive item in your budget, and it is mostly decaying feed which will pollute the water of your ponds and cause oxygen depletion and flareups of disease. It is vital to carefully determine the cheapest feed you can use and to use no more than the minimum amount.

As explained in the "Background on Aquaculture" section (Appendix), ideas on aquaculture feeding have undergone many changes in the last 30-40 years. Previous to the 1950s, cultured fish in ponds mostly ate their natural food organisms (sometimes phytoplankton, zooplankton, and aquatic plants, but more often the worms and insect larvae which live in the bottom mud). These natural food organisms are present in *Fish Farm* and can support small numbers of fish (e.g., a few hundred per hectare) without any input of food by you.

But if you wish to move beyond harvests of a few hundred kg/hectare and to profitable harvests of a few thousand kg/hectare (a kilogram, or kg, is 2.2 pounds; a hectare, or ha, is 10,000 square meters, or about 2.5 acres; 1 kg/ha is 0.88 pounds/acre), you must stock far larger numbers of fish than the bottom organisms can support. This will require feeding large amounts of pelleted artificial diet. These diets, which can be provided either in floating or sinking pellets, consist of ingedients such as soybean meal, fish meal, wheat, and corn. The floating pellets are more expensive, but they are commonly used because they allow the culturist to see how well the fish are feeding. The vigor of feeding is useful not only as a measure of fish health but also as an indication of how much feed to give.

The only decision which *Fish Farm* asks you to make about this feed is its protein content.

Protein is the most expensive part of fish feeds. In *Fish Farm*, a feed with 10% protein costs $0.21 per kg; a feed with 90% protein would cost $0.81 per kg. In addition, protein is a rather poor energy source and is broken down to release toxic ammonia, and this ammonia is converted to nitrate in the water with the consumption of large amounts of oxygen. Therefore, you should feed as little protein as possible. However, providing enough protein is vital for growth. Fish that are not fed a sufficiently high proportion of protein in their diets can compensate to a limited extent by eating more food, but this will cause more loading of organic matter into the pond and will lead to oxygen depletion.

The best protein policy is to feed the lowest percentage of protein which will cause the maximum growth rate. This varies with the type of fish. Carp can grow with 10-15% protein. The traditional figure for catfish is 32% (although catfish fry need 50% protein). And carnivorous trout need up to 60% protein in their feeds.

The **amount** of food fed to stock is usually given as the percent of their biomass which they eat per day. This varies with the type of fish, the water temperature (fish eat more in warmer waters), and the size of the fish (young, growing fish eat more). For example, grass carp (which eat hard-to-digest aquatic plants) eat 33% of their weight per day, *Tilapia* eat 15%, young carp eat 10%, and 1 kg carp only 2%. Catfish eat 3% when 25 g but 1.75% when 700 g.

Because of the expense of feed and its role in polluting the water, it should be your objective to feed the fish precisely the amount they wish to eat and not a milligram more. *Fish Farm* will do this for you automatically. If you choose to feed at a "varying" rather than at a constant rate, the program will change the feeding rate each day until it has attained a 90-99% feed consumption rate. A careful aquaculturist would do the same thing by observing the percentage of floating feed which the fish consumed and adjusting the feeding rate until almost all of the feed was eaten. The alternative, used for some of the tank experiments, is a constant feeding rate. In this case, the program feeds at the same rate, no matter how much the fish are eating or not eating.

The success of feeding in causing weight gain is measured by the FCR--the feed conversion ratio:

FCR = kg of feed given/kg of weight gained

*Therefore, the smaller the FCR is, the better.* FCRs less than 2.0 are considered good in catfish culture (and can be compared with FCRs of 7.5-20 for feedlot cattle). But if the fish are not growing even while much feed is being given, the FCR can get very large. If the FCR gets negative, it means the stock is actually losing weight.

Figure 3 shows the results of a protein content experiment similar to the one you will do.

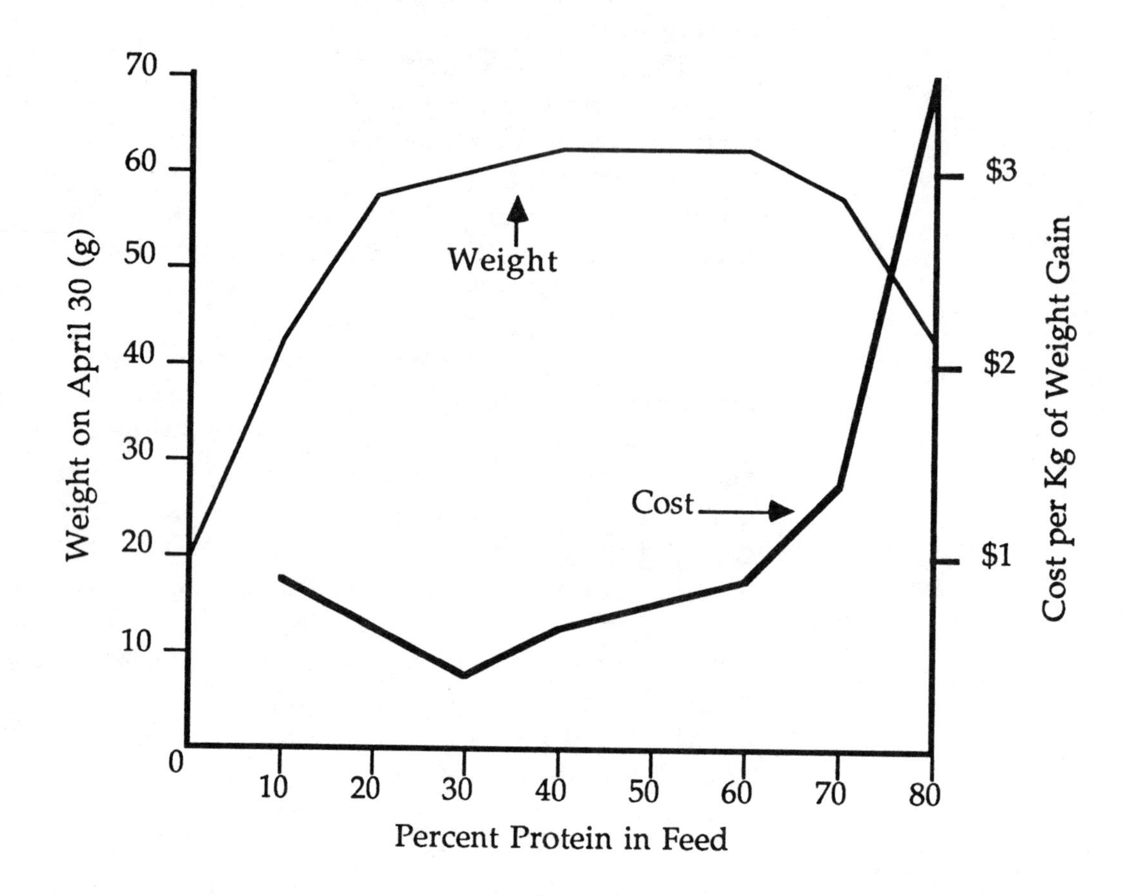


**Figure 3.** Growth of catfish as a function of protein content in the diet. Fish were grown in tanks at 25° C, 10 mg/l oxygen, and with a varying feeding rate.

## Objectives of the Tank Experiments

To summarize, there are three questions about your fish which your team must answer during the tank experiments:

1. Can the fish tolerate water temperatures of 25°-30° C without serious growth reduction? If so, it can be grown in static ponds without groundwater input.

2. How low an oxygen concentration can the fish tolerate

before its growth is reduced? These data are used to determine the "trigger concentration" for the mechanical aerators.

3. What feed protein percentage produces the most economical weight gain? This very important datum is used to determine the type of feed in all the pond experiments.

*Each student* should complete the worksheet on the next page.

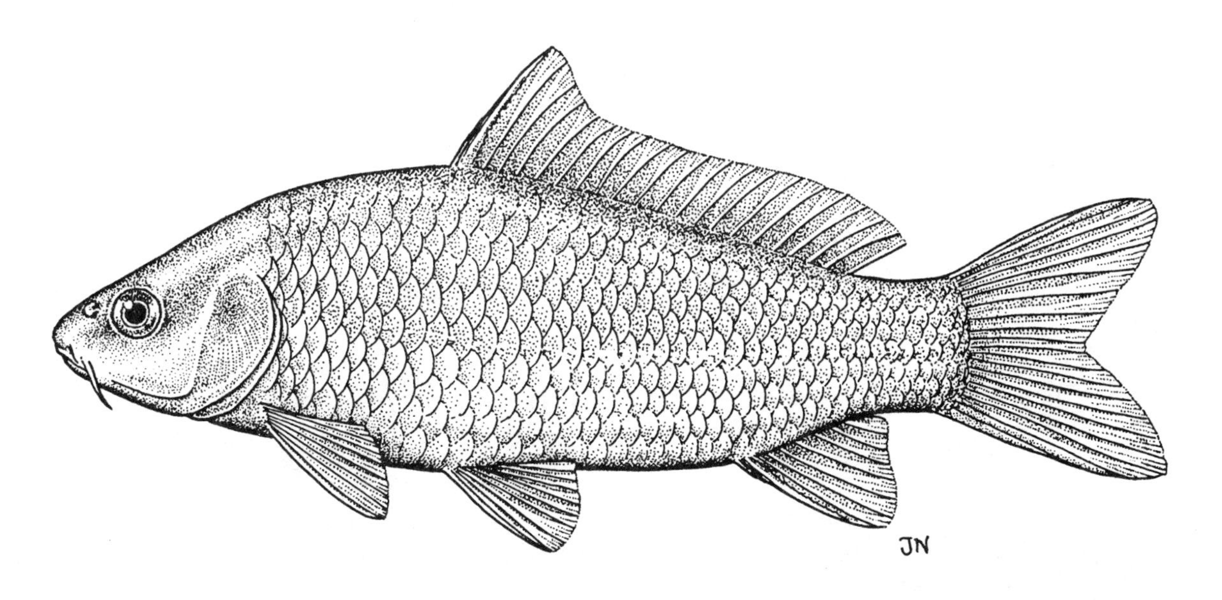


**Common Carp *(Cyprinus carpio)***

Carp are hardy bottom feeders which can thrive in crowded, organically polluted ponds. Carp is easily the most important cultured fish worldwide, accounting for almost 65% of the output of fish aquaculture. Most of this production comes from China, Russia, and Indonesia; in the U.S. the small carp industry is centered in Arkansas. But carp are widely regarded as "trash fish" in the U.S. American consumers complain that they have a "muddy" taste, and sportsmen dislike them because they root in bottom sediments and disrupt the life cycles of game fish.

## Worksheet 1

Name ______________________________

1. What results of the temperature experiment would convince you that you will need a rapid rate of groundwater input to raise your fish?

2. If your fish can grow without groundwater input, should you use it anyway? Why or why not?

3. If you intended to counter oxygen depletion by emergency aeration, how would you choose the oxygen concentration at which to begin aeration?

4. What is an "FCR"? Is it better if the FCR is large or small? Why?

5. Not feeding fish enough protein is obviously bad for nutritional reasons, but why is feeding them too much also bad? Give two reasons.

# Temperature Experiment Procedure

**Your Temperature Objectives:**

Collect growth and cost data at various temperatures and draw a graph similar to Figure 1 (page 11);

Find out if your fish can't tolerate temperatures of 25°-30° and therefore needs groundwater input.

**Note:** The details of versions of *Fish Farm* which run on Apple II, Macintosh, and IBM computers differ. When in doubt, follow the directions on the screen.

1. If your computer is off, skip to step 2. If your computer is still on the main menu after the conclusion of the catfish orientation exercise:

   a. **If you have an Apple II computer,** indicate that you wish to switch to tank experiments, and skip to step 4.

   b. **If you have an IBM or a Macintosh computer,** indicate that you wish to start again with a new fish. Then skip to step 4.

2. If your computer is off when you start, place your *Fish Farm* disk in the disk drive of the computer and turn the computer and monitor on.

3. After a title screen, you may be asked to adjust the program speed so a second timer ticks at the correct rate. Follow the directions on the screen.

4. **Apple IIe and Apple IIGS users only:** Indicate that you wish to work with **indoor tanks.** This question will come later for IBM and Macintosh users.

5. Select the fish (Fish A through Fish Z) which your instructor assigned to your team. Read the description,

and then accept the fish. Don't accept the wrong one, or you'll have to start all over again.

6. **IBM and Macintosh users:** Indicate that you wish to work with **indoor tanks** after your choice of fish.

7. You will experiment on temperature, but you must set the other conditions in the chamber for the first experiment. Set oxygen at 10 mg/l.

8. Following directions on the screen, set food to 50% protein, and use a **constant** feeding rate of **10%** of the tank's biomass per day.

9. Now **you** must select the temperatures you will use. All of the fish will have their optimum somewhere between 0° C and 40° C. **For a start**, list at least 5 temperatures in that range on the temperature data page (page 21). Select these temperatures with two considerations in mind:

   a. You should only use temperatures that are in the range of response of the fish. That is, if you selected 10°, 20°, 30°, 40° and 50° C and it turned out that your fish died of heat stress at 30° C, it would be a waste of time to continue through your list and try 40° and 50° C as well.

   b. Concentrate observations in "interesting areas." For example, in the graph showing the temperature response of the catfish (Figure 1, page 11), a sharp decline in growth is occurring between 30° and 35° C. It would be better to concentrate observations in this region than in the less interesting area between 10° and 20° C.

## Temperature Data for Fish _____

**Controlled Variables:**

Oxygen __________

Feed Protein __________

Feeding Rate __________

Constant or Varying? __________

| Temperature | Weight | FCR | Cost of Weight Gain | Comments |
|---|---|---|---|---|
| ________ | ________ | ________ | ________ | ________________ |
| ________ | ________ | ________ | ________ | ________________ |
| ________ | ________ | ________ | ________ | ________________ |
| ________ | ________ | ________ | ________ | ________________ |
| ________ | ________ | ________ | ________ | ________________ |
| ________ | ________ | ________ | ________ | ________________ |
| ________ | ________ | ________ | ________ | ________________ |
| ________ | ________ | ________ | ________ | ________________ |
| ________ | ________ | ________ | ________ | ________________ |
| ________ | ________ | ________ | ________ | ________________ |
| ________ | ________ | ________ | ________ | ________________ |
| ________ | ________ | ________ | ________ | ________________ |
| ________ | ________ | ________ | ________ | ________________ |
| ________ | ________ | ________ | ________ | ________________ |
| ________ | ________ | ________ | ________ | ________________ |

10. Enter your first temperature and then press the **ESC** key. This will start the experiment, and the computer will display the "digital display screen." A copy of the upper portion of this screen is shown on page 24.

| Fish A | 3/24 |
|---|---|
| Water temperature (C): | 15.0 |
| Dissolved oxygen (mg/L): | 10.0 |
| Eating (% of biomass): | 3.2 |
| Feed given (kg/hectare): | 1439 |
| Fish weight (g): 28 | |
| Surviving fish per tank: | 10 |
| Feed conversion ratio: | 3.9 |

The explanation:

a. **3/24** indicates March 24. The temperature and oxygen reading are the values you set.

b. **Eating** shows the food the fish ate on March 24, expressed as a percentage of their biomass. Remember that you are feeding 10%/day, so that in the example above they are eating less than 1/3 of what you are offering them.

c. **Feed Given** shows the feeding rate as kg/hectare.

d. **Feed Conversion Ratio,** or **FCR,** is the number of kg of dry fish food which causes each kg of fish fresh weight gain. The larger this number gets, the lower the efficiency of feed conversion. In catfish culture, values under 2 are considered good.

e. You may pause the simulation or go to the menu by following the directions on the screen.

**11.** The simulation will go **very** rapidly to April 30. Then it will automatically stop and revert to the harvest screen.

**12.** The harvest screen will tell you the average weight of the fish, the FCR, and the cost of feed per kg of weight gain. Record this information opposite the temperature on the temperature data page. If any fish died, record this fact in the "Comments" column. The program will ask you whether you wish to add this experiment to your database. If it was a valid experiment (and not the result of an error in entering conditions or some other problem), add the results to the database. Otherwise, discard the results. Then indicate that you wish to **continue to the main menu**. Finally, indicate that you wish to **do another** experiment. Select **indoor tanks**.

**13.** Repeat steps 10-12 for the other temperatures you wish to test. *After you have found the approximate range of your fish's tolerance, test more temperature values* to establish your fish's performance over a wide temperature range, and be sure to test its tolerance for warm temperatures (around 30° C). Because the same temperature may give different results in different simulations, do at least two simulations at important temperatures. To be sure you have attained both objectives cited in the "temperature" box above, select **review past tank experiments** from the menu at the end of your last experiment. Then select **Cost of Weight Gain (y) vs. Temperature (x)** as the graph. Ask your instructor to look at your results.

Using your temperature data and the data on temperature conditions in the ponds (Table 1, page 11), determine whether the fish cannot tolerate temperatures much higher than 20° C or can grow well only in a narrow temperature range around 18° C. If so, plan on using rapid groundwater in the stocking density experiment and the production run.

If the fish seems to tolerate temperatures from 10° to 30° C, there is no reason to use groundwater input. This would be good news because it allows you to use the full 200 hectares of ponds to produce your fish.

# Oxygen Experiment Procedure

> **Your Oxygen Objectives:**
>
> Collect growth and cost data at various oxygen concentrations and draw a graph similar to Figure 2 (page 15).
>
> Determine the **lowest** oxygen concentration which your fish can tolerate without serious growth reduction.

1. If you have just completed the temperature experiment and your computer is still running *Fish Farm*, skip to step 2. If not, perform steps 1, 2, 3 and 4 on page 21, then start with step 3 below.

2. If you have just completed the temperature experiment, **clear the tank database** (option 3 on the main menu).

3. Indicate that you wish to perform **another experiment,** and then choose **indoor tanks.**

4. Choose a non-stressful temperature which produces a good weight gain (20° should work for most fish). Set the temperature to this value. Then set the feeding regime to **50% protein** feed, and a **constant** feeding rate of **10%** of biomass per day.

5. Determine at least five oxygen values which you feel will show the response of your fish to low oxygen conditions. List these on the oxygen data page (page 27).

6. Enter your first oxygen value and then press the **ESC** key. This will start the experiment, and the computer will display the "digital display screen." An explanation of this screen is on page 24.

7. Pressing **ESC** will bring you back to the menu from which you chose the tank conditions. Follow the directions on the screen to pause the simulation and to restart it.

# Oxygen Data for Fish _____

**Controlled Variables:**

Temperature __________

Feed Protein __________

Feeding Rate __________

Constant or Varying? __________

| Oxygen | Weight | FCR | Cost of Weight Gain | Comments |
|---|---|---|---|---|
| | | | | |
| | | | | |
| | | | | |
| | | | | |
| | | | | |
| | | | | |
| | | | | |
| | | | | |
| | | | | |
| | | | | |
| | | | | |
| | | | | |
| | | | | |
| | | | | |
| | | | | |

8. The simulation will go **very** rapidly to April 30. Then it will automatically stop and revert to the harvest screen.

9. The harvest screen will tell you the average weight of the fish, the FCR, the cost of feed per kg of weight gain, and how many fish died. Record this information

opposite the oxygen level on the oxygen data page. If any fish died, record this fact in the "Comments" column. The program will ask you whether you wish to add this experiment to your database. If it was a valid experiment (and not the result of an error in entering conditions or some other problem), follow the directions on the screen to add the results to the database. Otherwise, discard the results. Then indicate that you wish to **continue to the next experiment**. Finally, indicate that you wish to **start another experiment** and select **indoor tanks**.

10. Repeat steps 6-9 for your other oxygen values. Then collect additional oxygen observations to precisely determine the oxygen concentration *at which your fish first begins to grow poorly*. This is the oxygen value which will trigger emergency aeration during the stocking density exercise. To be sure you have attained both objectives cited in the "oxygen" box above, select **review past tank experiments** from the menu. Then select **Cost of Weight Gain (y) vs. Oxygen (x)** as the graph. Ask your instructor to look at your results.

## Feeding Experiment Procedure

**Your Feeding Objectives:**

Collect growth and cost data at various feed protein contents and draw a graph similar to Figure 3 (page 18).

Determine the feed protein percentage which allows your fish to gain weight with the lowest cost per kg of weight gain.

1. If you have just completed the oxygen experiment and your computer is still running *Fish Farm*, skip to step 2. If not, perform steps 1, 2, 3, and 4 on page 21, then start with step 3 below.

2. If you have just completed the oxygen experiment, **clear the tank database** (option 3 on the main menu).

3. Indicate that you wish to **perform another experiment,** and then choose **indoor tanks.**

4. First, you must set tank conditions for the feeding experiment. You wish to set these at values which are not stressful so that the fish are responding only to feeding. Most fish will do well at 20° C and with 10 mg/l of dissolved oxygen.

5. **For a start,** select at least five values for percent protein of the feed and list them on the feeding data page (page 28). They should be fairly evenly spaced between 0% and 100% protein.

6. Select **feeding** from the menu and enter the first percent protein you choose into the program. Indicate that you wish the feeding rate to **vary.**

7. Enter your first protein value and then press the **ESC** key. This will start the experiment, and the computer will display the "digital display screen." An explanation of this screen is on page 24.

8. Pressing **ESC** will bring you back to the menu from which you chose the tank conditions. Follow directions on the screen to pause and restart the simulation.

9. The simulation will go **very** rapidly to April 30. Then it will automatically stop and revert to the harvest screen.

10. The harvest screen will tell you the average weight of the fish, the FCR, the cost of feed per kg of weight gain, and how many fish died. Record this information opposite the protein level on the feeding data page. The program will ask you whether you wish to add this experiment to your database. If it was a valid experiment (and not the result of an error in entering conditions or some other problem), add the results to the database. Otherwise, discard the results. Then indicate that you wish to **continue to the next experiment.** Finally, indicate that you wish to **start another experiment.** Select **indoor tanks.**

## Feeding Data for Fish ______

**Controlled Variables:**

Temperature ______

Oxygen ______

Feeding Rate ______

Constant or Varying? ______

| Protein | Weight | FCR | Cost of Weight Gain | Comments |
|---|---|---|---|---|
| | | | | |
| | | | | |
| | | | | |
| | | | | |
| | | | | |
| | | | | |
| | | | | |
| | | | | |
| | | | | |
| | | | | |
| | | | | |
| | | | | |
| | | | | |
| | | | | |
| | | | | |
| | | | | |
| | | | | |
| | | | | |
| | | | | |

**11.** Repeat steps 7-10 for the other protein percentages you wish to test.

**12.** When you have tested your first group of protein percentages, select "Review your past tank experiments" from the menu. Then select "Cost of Weight Gain (y) vs. Feed Protein (x)" as the graph. Inspect your data and decide which type of feed will give the most economical weight gain (that is, the lowest cost per kg of weight gain). *Then test 5-6 more percentages around this optimum value* to determine the optimum more exactly. Determining optimum protein is very important for profitable pond experiments, and the only way to do it accurately is to use plenty of replicates.

**13.** To be sure you have attained the objective cited in the "feeding" box above, select "Review your past tank experiments" from the menu, select "Cost of Weight Gain (y) vs. Feed Protein (x)" as the graph, and ask your instructor to look at your results.

When you are finished with the tank experiments, you should have determined:

a. the optimum temperature for your fish.

b. the range of temperature over which your fish can grow well.

c. the oxygen level at which your fish begins to grow poorly and die.

d. the percent protein in the feed which causes weight gain at the cheapest cost per kilogram.

INSTRUCTOR CHAPTER 3

# The Second *Fish Farm* Lab

The objectives of the second *Fish Farm* lab are to:

1. Give the students some advice on writing the *Fish Farm* report (20 minutes);
2. Go over the data presentation worksheets (Worksheet 2) (15 minutes);
3. If separate groups did the temperature, oxygen, and feeding experiments in Lab 1, have these separate groups present their tank experiment data so that the whole class agrees on groundwater, oxygen, and protein for the unknown fish (30 minutes);
4. Introduce and do the stocking density exercise (80 minutes);
5. Do the production run exercise (20 minutes).

**Before Class**

The students should have read Chapters 3 and 4 in the *Fish Farm* manual (pages 35-64) and completed Worksheets 2 (page 49) and 3 (page 65).

## The Final Report

1. Give the deadline date for the final report. This report may be scary to the students because it has raw data pages, tables, graphs, and a written portion. So a few words of advice (delivered at the beginning when attention is high and impatience is still low) may be in order.
2. Tell the students to turn to page 77 in their student manuals. Almost everything about the report is in Chapter 5, but some of the students may never see it unless you draw their attention to it.

3. Tell the students that there are four parts to the report:

   a. a cover sheet (page 78 of the student manual);

   b. the raw data pages removed from the student manual (pages 79-81 of the student manual);

   c. the tables and graphs which formally present their *summarized* data on sheets of graph paper attached to their discussion (pages 81-83 of the student manual);

   d. the discussion and conclusions, *typed* 1-2 pages (double-spaced) which summarize their results, point out weaknesses, and give their culturing recommendations (pages 83-87 of the student manual).

4. Have the students turn to the different pages of Chapter 5 of the student manual as you point out the major features of these report parts. Illustrate your remarks with the example of the protein experiment, which the manual takes from the raw data stage through the discussion stage.

5. Draw the students' attention to pages 87-89 of the student manual and summarize how the reports will be graded.

## The Data Presentation Worksheet

6. Collect the data presentation worksheets, but don't collect the graphs (which should be on a separate sheet of graph paper).

7. Explain to the students that teaching them how to present scientific data clearly and correctly is one of the major goals of any science lab. Review the high points of Chapter 3, and simultaneously go over the answers to Worksheet 2 (next page). The "high points" include:

   a. dependent, independent, and controlled variables;

   b. when to use tables (which will get you into discrete as opposed to continuous variables);

c. the basics of table construction, with emphasis on informative captions that mention the controlled variables in the experiment;

d. how to make a line graph, with emphasis on properly labeled axes and informative captions;

e. bar graphs, with emphasis on when to use a bar graph instead of a line graph;

f. frequency histograms, with emphasis on the fact that the x-axis shows discrete intervals rather than continuous values, and that the y-axis shows the number of occurrences in each of the intervals;

g. how to present replicates in both tables and graphs.

An excellent way to answer worksheet questions 4 and 5 is to find a student with a good response and have that student present his/her response to the class, either on an overhead transparency or the blackboard. If each team did its own temperature, oxygen, and protein experiments, collect the graphs after this presentation and skip to step 13. If each team did only one of these experiments, proceed with step 8.

## Worksheet 2 Discussion Guide

**1.** The dependent variables are all the variables which respond to temperature, oxygen, or feed protein. Some of the more prominent dependent variables are fish weight, the number of fish surviving and cost per kilogram of weight gain. The only independent variables are temperature in the temperature experiment, oxygen in the oxygen experiment, and protein in the protein experiment. Some of the controlled variables (in the temperature experiment, for example) are oxygen concentration, protein content and the feeding rate (and also implied variables like the number of fish per tank and the type of fish used).

**2.** See answer to question 1.

**3.** Type of fish is the only discrete variable. The rest are continuous.

### Worksheet 2 Discussion Guide

4. Because both independent variables (fish type and team place) are discrete, plus the fact that an unrelated variable like the percentage of teams not making money is also thrown in, a table is the best solution:

**Table 1.** Five-year profits for the most profitable teams, and percentage of teams losing money, for Fish A, B and C during the fall, 1992 semester.

| | Fish A | Fish B | Fish C |
|---|---|---|---|
| First Place Team | $1,980,000 | $410,000 | $1,010,000 |
| Second Place Team | $1,920,000 | $370,000 | $920,000 |
| Third Place Team | $1,870,000 | $320,000 | $890,000 |
| Percent Losing Money | 0% | 12% | 3% |

5. Because temperature, oxygen, and feed protein are continuous variables and hopefully there are a large number of data points, a line graph is the best solution. See Figures 1-3 on pages 13-18 of the student manual for sample presentations. Note that the students will need a caption similar to the ones in those figures in order to complete the assignment.

## Sharing of Tank Results

Skip this section if each team did its own temperature, oxygen, and feeding experiments. In that case, go to step 13.

8. For practical application, tell the students that now they have to present data to each other: the class as a whole has to decide whether the unknown fish needs groundwater, at which oxygen concentration the aerators need to come on, and what protein content the feed should have. And the class will only make its decision after looking at the data.

9. If you will be using an overhead projector, pass out transparencies and pens to each team and have them put their temperature, oxygen, or protein graph on it. They should be able to do this quickly by tracing it from their graphs that they still have. If you will be using the blackboard, have each team send a representative to a different part of the board and present the graphs that way.

10. Have each team present their results to the class, with temperature first, then oxygen, and then protein. Presentations will probably be hesitant and short, but they should be more than, "OK, here's the graph we got." First they must indicate how the curve *should* look (perhaps by referring to the transparency of the relevant graph from the manual). This will get students used to prediction as a part of any results presentation. Then they should use their results to answer the question that concerns their team (the three questions are listed on pages 18-19 of the student manual). If they don't offer an answer, ask them for it. Ask if the class agrees. If you glance at your "*Fish Farm* Fish" summary sheet and see that they're presenting the wrong conclusion, don't tell them. They'll find out in the next hour or so.

11. After the last team has presented data, write the *accepted* (not necessarily correct) answers to the three questions on the overhead so that no one is left behind.

12. Make sure that all the students put their names on the graphs, and collect them.

## The Stocking Density Exercise

13. Collect the stocking density worksheets (Worksheet 3, page 65) and go over the answers (next page). This will be a good review of the material in Chapter 5.

14. Use the transparencies to continue the review of the chapter. It's especially worth drawing attention to Figure 14 (page 54 of the student manual), which shows the range of profits from catfish experiments. As stocking density goes up, profits become higher and **more variable**.

## Worksheet 3 Discussion Guide

1. At too low a density, the fish will grow well, but there will not be enough product to cover fixed costs. The more expensive taxes, land rental, vehicle registration fees, and so forth are, the more fish the culturist **must** raise to make a profit. At too high a density, the fish are stressed by oxygen depletion and other factors, have a high FCR, and may die. Death is obviously bad, but the high FCR greatly reduces profits because it wastes tons of expensive feed and simultaneously cuts down on the crop.

2. Disease in *Fish Farm* is caused by a) high bacterial populations, and b) fish stress (from temperature, oxygen depletion, excess ammonia, or social strife). Stress reduces the efficiency of the fish immune system. To counter disease, reduce fish stress, cut down the feeding rate (which will reduce bacterial populations), and give antibiotics. Antibiotics will reduce disease deaths, but only for a few days.

3. Groundwater a) brings in oxygen (9.5 mg/l), b) keeps temperatures low (18°) so the water holds more oxygen, c) sweeps away ammonia and decaying food, and d) sweeps away bacteria which might cause or spread disease. However, groundwater will be ineffective in preventing social strife since this problem is caused by high population density, not poor water quality.

## Worksheet 3 Discussion Guide

4. The protein experiment sometimes yields a slightly incorrect value for the minimum protein concentration. This happens even if the students do the experiment correctly and have adequate replication. However profit in the stocking density experiments is a very sensitive indicator of correct protein. The "arbitrary shift" experiments (page 63 of the student manual) use the stocking density experiments to verify correct protein percentage.

5. For static pond carp and catfish, a stocking density of 2,000 - 10,000 fingerlings per hectare seems reasonable. For raceways, it all depends on the water flow rate. If the water is exchanged many times per day, you can stock millions of fish (depending on their size). For the more modest water flow rates in *Fish Farm*, perhaps 200,000 **catfish** per hectare would be a good density, but of course this will vary for the unknown fish.

**15.** Another point is made by the transparencies of Figures 15 and 16 (pages 55 and 56 of the student manual). Stocking density experiments *will* yield a stocking density curve which has a peak in the middle and falls off on either end. But variability and a small range of attempted stocking densities may give what looks like a random scatter of points. Replication and a large range of stocking densities are necessary to see the true pattern.

**16.** Draw the students' attention to the troubleshooting section on page 61 of the student manual. This should allow students to diagnose their own problems.

**17.** Make sure that each team has the right fish, and let them start the computers. Students should follow the directions on page 66 of the student manual. The objective of maximizing profit will need little explanation.

**18.** As they begin, remind the students again that their lab notebook records will be 30% of the report grade. Page 87 of the student manual summarizes the characteristics of a good data page and the points attached to each characteristic.

**19.** There may be dissatisfaction when a team just can't make a profit. Usually the reason will be either an incorrect protein percentage or an incorrect groundwater regime:

**a.** Not using groundwater when it is needed is easy to spot. The fish will be growing fine when the water is cool but will die spectacularly when the water gets warm in late April or early May. "Heat stress" or "rapid temperature change" will be listed as the cause of death.

**b.** Not using **enough** groundwater is a common problem. Groundwater is best either at 0% input/day or 190% input/day. Certainly, any input which is less than 100%/day is probably not doing much for the waste and oxygen situation, and at the same time is drastically reducing the size of the pond.

**c.** Using a wrong protein percentage (usually too high) is indicated when the fish seem to be growing well but the farm just can't make a profit.

**d.** The given method of estimating the correct protein percentage with the cost per kilogram of weight gain works, but is not perfect. It is worthwhile for everyone to make an arbitrary shift in protein percentage (say up 3% and then down 3%) from their estimated optimum protein, just to make sure they're really using the best protein content possible. This is incorporated into step 23 in the directions and is explained on page 63.

Even though the class may have accepted oxygen, protein, and groundwater conclusions at the beginning of the lab, a team can change any of these parameters if they think they can earn a higher profit by doing so. *However, they have to state a reason for the change in their report.*

**20.** The stocking density experiment also requires them to determine time of harvest. This is explained on pages 52 and 53 of the student manual. In 24 fish out of 26, the time of harvest is on November 5 at the end of the season (called "ES" in the program). Only Fish A and Fish Y should be harvested when they attain marketable weight ("MW").

**21.** Let teams work until they think they are ready for the production run.

## The Production Run Exercise

**22.** This is just a short repeat of the stocking density experiments, but the important rule is that it must be five **consecutive** pond simulations. Teams can change conditions between simulations if they want, but once they start, they must do five consecutive runs.

**23.** Give the production run record (blank form in Appendix B) to each team, and tell them to fill it out as they do the production run. Collect the production run records, be sure that the five-year profit adds up correctly, find out who made the most money, and post this team's five-year profit and culturing conditions on an overhead or the blackboard. This will allow every team to compare their results with those of the "winners."

**24.** Warn the students to record their production run data on the the production run data page, not only on the production run records you give them.

**25.** Remind the students of the due date for the *Fish Farm* report.

**Homework**

Chapter 5 of the *Fish Farm* manual gives advice on completion of the final report. And, of course, the major homework from this lab is the final report.

FISH FARM

# Lab 2

CHAPTER 3

# Scientific Data Presentation

Scientific conclusions depend on data, and conclusions can only become widely accepted by other scientists if the data can be clearly and concisely communicated. Therefore, you have to learn a few of the common ways by which scientists present their results.

First, "data" is a plural word (the singular is *datum*, which is Latin for "something given"). Therefore, "These data *are* incomplete," is correct.

Every data presentation should strive to meet three ideals:

*Appropriateness of format.* Several different kinds of tables and graphs might be used to present a data set, but usually only one way is the best way.

*Clarity.* There should be no doubt about what each data point means. The methods used to produce it should also be clear. Correctly labeling graphs and tables is the important step to achieve clarity.

*Conciseness.* The data should be presented in a summarized, rapidly understandable form.

## Variables

A variable can be either an experimental condition or an experimental response. In any experiment, there are three types of variables.

The **dependent variable** is the response being studied. For example, if we are trying to determine the factors affecting profits in an aquaculture experiment, profit is the dependent variable, the one which depends on (or responds to) the various experimental manipulations we are performing.

The **independent variable** is the variable which is affecting the dependent variable, and *which is being systematically varied in the current experiment.* Continuing with the example above, say we are investigating the effect of stocking density (the number of fish put in a pond) on profit. Stocking density is the independent variable, and profit is the dependent variable.

The **controlled variables** are factors which might be affecting the dependent variable, but which are being held constant in this experiment. In the example above, type of food, temperature, groundwater input and so forth all could affect profit, but they are being held constant so that we can clearly determine the effect of stocking density on profit. Values of the controlled variables are usually listed in the caption of the table or graph so that the reader can tell the full conditions of the experiment.

The two main means of presenting the variables are tables and graphs.

## Tables

*Any* scientific data *can* be presented in a table, but students tend to use tables when they should use graphs (perhaps because tables are easier to construct). First, use a table when you have very few data points to present. The data in the table below would not warrant a graph because there are not enough points to draw any conclusions about a smooth relationship between stocking density and yield.

**Table 3.** Yields in pond experiments raising catfish at three different stocking densities. All experiments used 30% protein feed, no groundwater input, and aeration when oxygen reached 5 mg/l.

| Fish per Hectare | Yield (kg/hectare) |
|---|---|
| 5,000 | 2,892 |
| 10,000 | 4,280 |
| 15,000 | 5,100 |

Despite the fact that the table above is small, note that it still must have a table number, a descriptive title, and a labeling of all its columns with their units of measurement (e.g., kg/hectare). Note that the independent variable (stocking density) is presented on the left, the dependent variable on the right, and the values of the controlled variables (like feed protein percentage) are in the table title.

Another situation which makes a table preferable to a graph is when several dependent variables are being reported at once. For example:

**Table 4.** FCR's, mortality rates, and yields in pond experiments raising catfish at three different stocking densities. All experiments used 30% protein feed, no groundwater input, and aeration when oxygen reached 5 mg/l.

| Fish per Hectare | FCR | % Mortality | Yield (kg/hectare) |
|---|---|---|---|
| 5,000 | 1.8 | 0.0 | 2,892 |
| 10,000 | 2.1 | 0.8 | 4,280 |
| 15,000 | 2.6 | 4.9 | 5,100 |

This table shows that there could be many dependent variables in one experiment. For any experiment involving some number of fish per hectare (the independent variable), responses in addition to the ones above might be the average weight of the fish, the profit, the number of times aeration had to be used, disease incidence, and so on. Because they are all responding to the independent variable, they are **all** dependent variables.

The table above had one independent variable and several dependent variables, but you might have to present data where there are several **independent** variables. Say we were growing two kinds of catfish (strain A and strain B), both with and without antibiotic added to the feed, and are monitoring yield. For this simple situation, we might use a table which looks like this:

**Table 5.** Yields (kg/hectare) of strain A and strain B with and without antibiotic added to feed. All experiments used 7,500 fish/hectare, 30% protein feed, no groundwater input, and aeration when oxygen reached 5 mg/l.

| | Strain A | Strain B |
|---|---|---|
| With antibiotic | 3,900 | 3,200 |
| Without antibiotic | 3,300 | 3,188 |

A more complex table with subheadings would be required if we were doing the comparisons above, but over a range of stocking densities instead of just at 7,500 fish/hectare:

**Table 6.** Yields (kg/hectare) at several stocking densities (fish/hectare) of strain A and strain B with and without antibiotic (Anti) added to feed. All experiments used 30% protein feed, no ground-water input, and aeration when oxygen reached 5 mg/l.

| Stocking Density (fish/hectare) | Strain A | | Strain B | |
|---|---|---|---|---|
| | With Anti | Without Anti | With Anti | Without Anti |
| 5,000 | 2,480 | 2,216 | 2,150 | 2,180 |
| 10,000 | 4,600 | 4,194 | 4,215 | 4,090 |
| 15,000 | 7,019 | 6,445 | 6,455 | 6,330 |

The last situation when a table is usually better than a graph is when the independent variable is discrete rather than continuous. A variable is **continuous** when it varies smoothly between values. For example, temperature is a continuous variable because for any two temperatures, there are an infinite number of intermediate temperatures in between. Other examples of continuous variables are fish weight, feed protein percentage, stocking density, and time. On the other hand, a discrete variable can assume only a limited number of "discrete" or separate values. Examples would be sex (either male or female), states of the USA (only 50 possible choices), or yes-or-no experimental treatments like "Feed contained antibiotics at the recommended rate" and "Feed contained no antibiotics."

However, this last example shows how recommendations can be difficult in certain situations. The "yes/no" antibiotic example is discrete and best handled by a table, but it could be made continuous (and require a graph) if we decided, for example, to give antibiotics at 0%, 25%, 50%, 75%, and 100% of the recommended dose. Also, although states are discrete variables, a table which shows hectares of aquaculture ponds in a group of states could also be presented as a bar graph (see page 42).

# Graphs

Graphs are data "pictures" and have great power to communicate data patterns rapidly and efficiently. The three major types of graphs which are used in scientific presentations are line graphs, bar graphs and frequency histograms.

## Line Graphs

These are the "workhorses" of data presentation, so you should learn how to do them correctly. Use a line graph when you have more than 3 or 4 data points, the independent variable is continuous, and you only wish to portray 1 or 2 dependent variables.

Let's say that the students who prepared Tables 3 and 4 did some additional simulations at other stocking densities, including some duplicate simulations at the same stocking densities. They might present their data in a table:

**Table 7.** Average fish weights, mortality rates and yields (kg/hectare) in pond experiments raising catfish at different stocking densities. All experiments used 30% protein feed, no groundwater input, and aeration when oxygen reached 5 mg/l.

| Fish per Hectare | Weight (g) | % Mortality | Yield |
|---|---|---|---|
| 500 | 660 | 0.0 | 330 |
| 1,000 | 702 | 0.0 | 702 |
| 2,500 | 660 | 0.0 | 1644 |
| 5,000 | 630, 610 | 0.0, 0.0 | 2892, 2690 |
| 7,500 | 605, 590 | 0.0, 0.0 | 4540, 4425 |
| 10,000 | 576, 520 | 0.0, 0.0 | 5769, 5200 |
| 15,000 | 478 | 0.0 | 7183 |
| 20,000 | 412 | 1.8 | 8092 |
| 25,000 | 347 | 19.9 | 6740 |

The obvious message of this table is that overstocking produces a decline in fish weight, growing mortality, and yield which rises at first but then drops off. But the data are more suited for a line graph because there are many data points and stocking density is a continuous variable. If your major interest is presenting the yield information, you would do it like this:

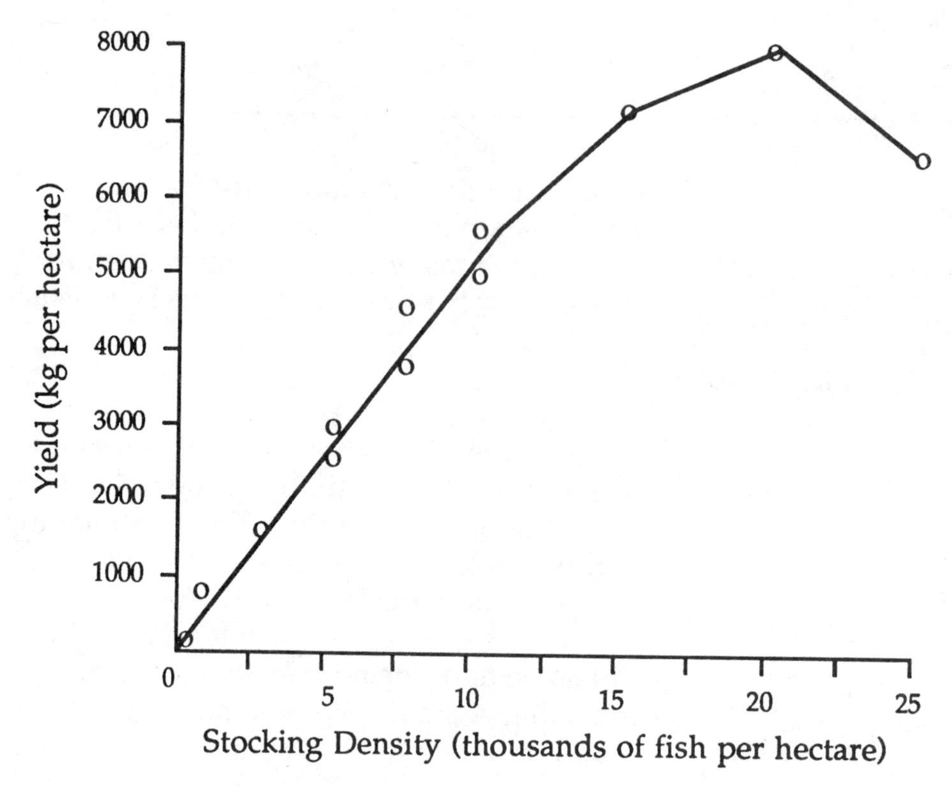


**Figure 4.** Yield of catfish from outdoor ponds using 30% protein food, no groundwater input, and aeration when dissolved oxygen reached 5 mg/l.

Note the characterisitics of a good line graph:

1. The independent variable is on the x-axis (the horizontal axis) and the dependent variable is on the y-axis. Don't get this mixed up! If you plotted stocking density on the y-axis, you would be telling your audience that you think stocking density is determined by yield, and not the other way around.

2. The axes are clearly labeled and have units of measurement (e.g., kg/hectare).

3. The axes have intervals which are all the same size. Every interval on the x-axis represents 2,500 fish/hectare; every interval on the y-axis represents 1,000 kg/hectare of yield.

4. The curve passes through the "cloud" of points rather than connecting points exactly. The purpose of the curve is to illustrate the general trend of the data, not emphasize its variability.

5. The data points (o's) are present and easy to see (not untidy, barely visible dots).

6. The figure caption is numbered and (like the table titles) describes the conditions of the experiment.

7. The curve takes up the whole area of the graph.

Sometimes it's a good idea to present two dependent variables on one graph, especially when the two curves are not likely to be confused. In the example above, the obvious candidate is individual fish weight, which declines as the fish become more and more crowded. A good way to handle this is to put another y-axis on the right side of the graph:

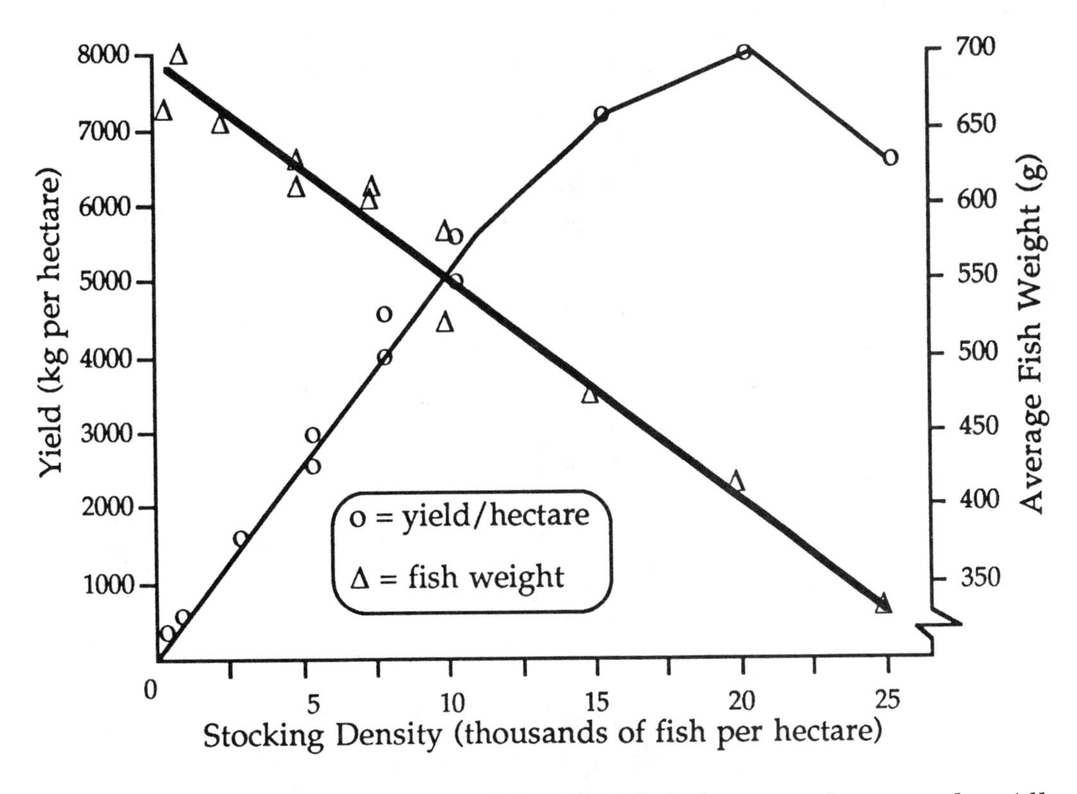


**Figure 5.** Yield and average weight of catfish from outdoor ponds. All experiments used 30% protein food, no groundwater input, and aeration when dissolved oxygen reached 5 mg/l.

Note that weight points and curve are distinguishable from the yield points and curve. Also, in order to allow the weight curve to take up the full area of the graph, the weight axis was truncated (it only extends from 350-700 g, not 0-700 g).

But this is pointed out to the reader by the zig-zag portion (which indicates a break) at the bottom of the weight axis.

As a rule, you should not plot more than two dependent variables on one line graph because the result will probably be a busy, confusing presentation.

## Bar Graphs

Bar graphs should be used when the independent variable is discrete rather than continuous, there are more than three categories of the independent variable, and there is some reason why a graph rather than a table is desirable. For example, consider Figure 6 (which contains fabricated data and is being used solely for illustration):

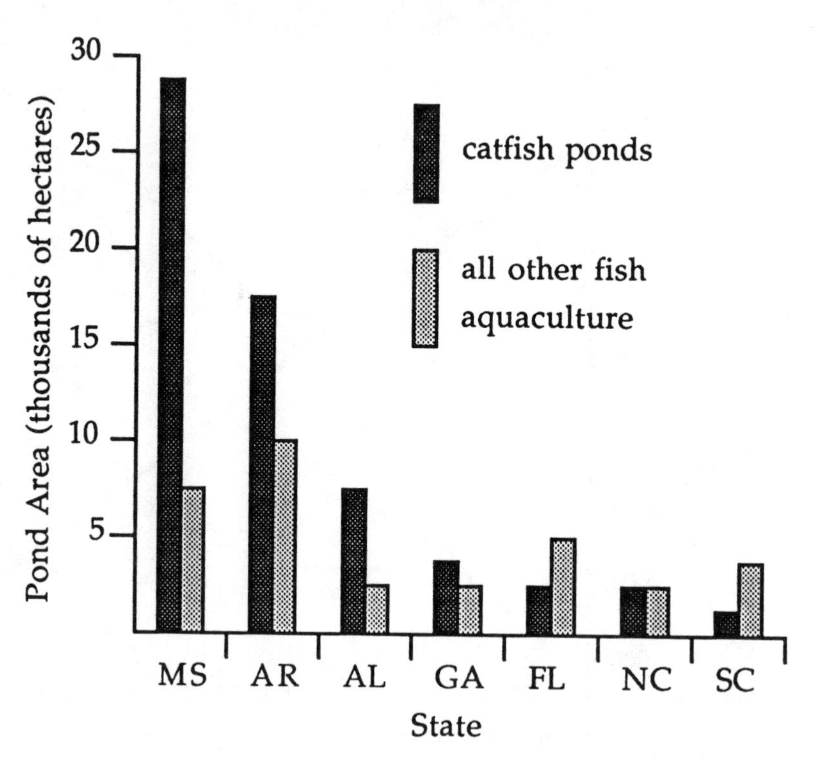


**Figure 6.** Pond area devoted to catfish aquaculture and culture of other fish in several southern states, 1985.

The states here are classic discrete variables. Note that the bars referring to the different states are well separated from each other, and the two bars for each state are distinguishable by shading. The chief advantage of a bar graph over a table in this case is that the bar graph gives visual impact to the dominance of Mississippi, and this impact is strengthened by the fact that the states are arranged in order of their catfish pond area. This arrangement is a good technique when the objective is emphasis of the most important members of a group.

## Frequency Histograms

A frequency histogram is a type of bar graph which shows a plot of the numbers of observations which fell into several categories of some measurement. It is used to show the variability in a data set. An example which will be familiar to all students is the "bell-shaped curve" usually seen in grade distributions:

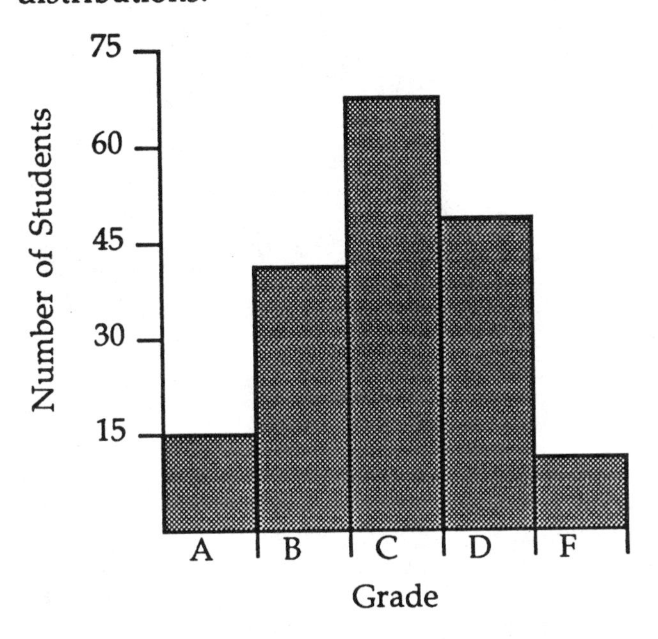


**Figure 7.** The distribution of final grades in Professor X's biology lecture section, Fall 1991.

Note that the x-axis shows grades, which are nonoverlapping categories (that is, a single student can only have one grade). The tic marks on the x-axis are just dividers which separate one grade from another. The y-axis contains the number of students who received each grade. Thus 15 students received A's and 50 students received D's.

If Professor X wanted a more informative distribution, he could make the x-axis a series of nonoverlapping 5-point intervals. This would allow examination of the widely varying students who received F's:

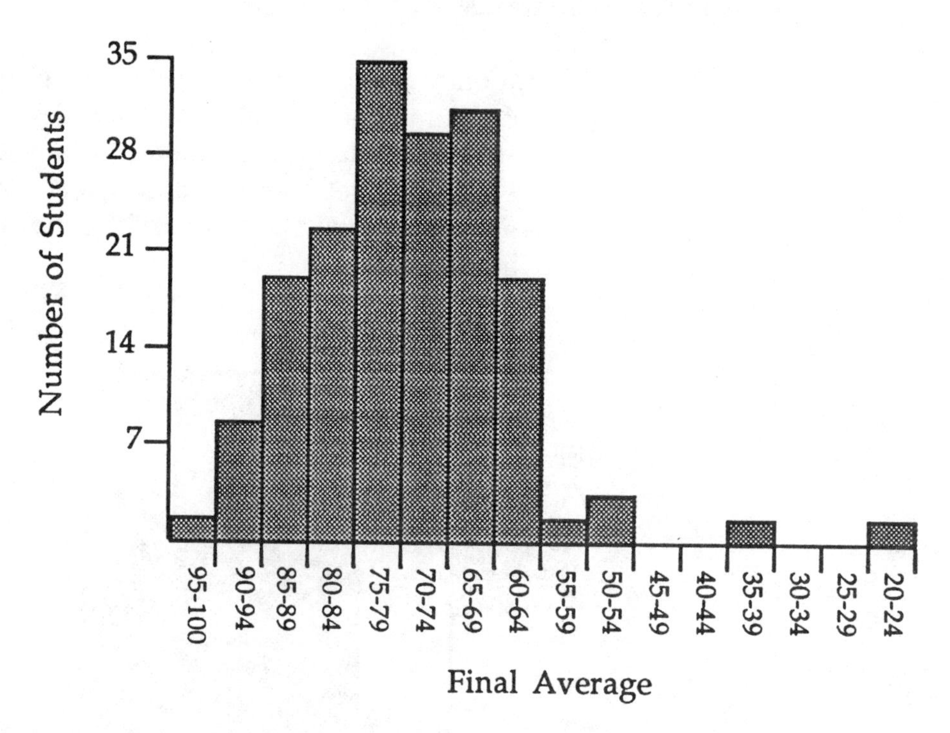


**Figure 8.** Distribution of final averages in Professor X's biology lecture section, Fall 1991.

Again, the tic marks on the x-axis are dividers. The interval designations (such as "95-100") are in the middle of the interval, **not** on the tic marks. The y-axis here still shows the number of students in each category, but the numbers are smaller than in Figure 7 because there are fewer students in these smaller intervals. Finally, note that some intervals are empty--no students received a grade from 45% to 49%, for example. This is a perfectly valid result.

A final use of a this type of graph is to compare two distributions. Let's say Professor X decides to compare the grades of biology majors and students who are not biology majors. One way to do this is to plot a bar for each major category within each grade interval:

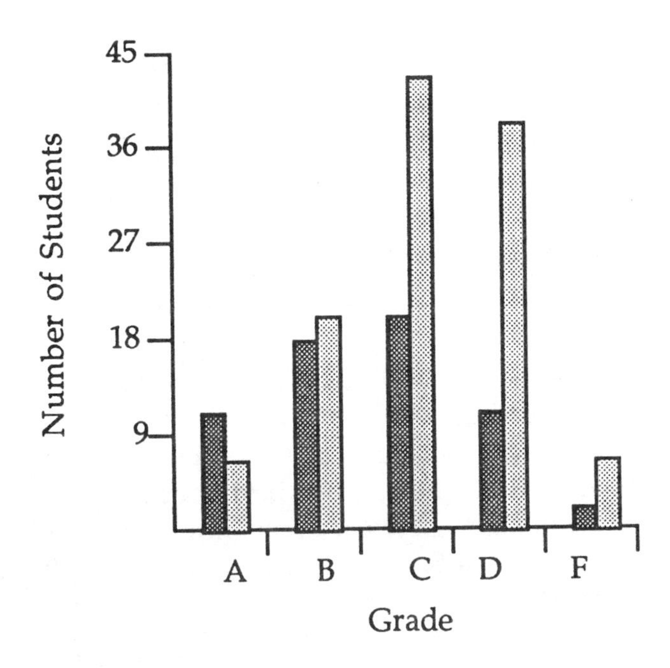


**Figure 9.** Distribution of course grades for biology majors (dark bars) and non-biology majors (light bars) in Professor X's lecture section, Fall 1991.

Professor X could also plot **percentages** of each group (rather than numbers of students) on the y-axis. For example 16% of the biology majors (and 4% of the non-biology majors) received an A:

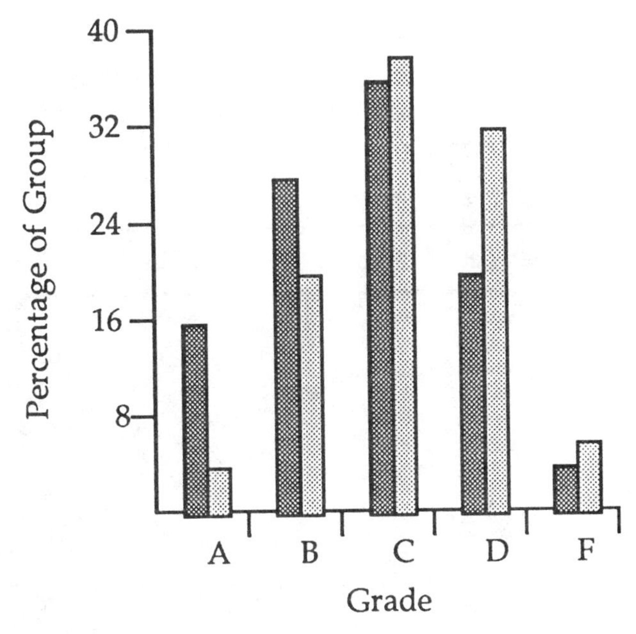


**Figure 10.** The percentage of biology majors (dark bars) and non-biology majors (light bars) who received A's, B's, C's, D's, and F's in Professor X's biology lecture section, Fall 1991.

Figure 10 makes it clear that biology majors had a greater chance of receiving an A or a B from Professor X, and a smaller chance of receiving a C, D, or F.

## Presenting Replicates

Replication (taking more than one observation at each value of the independent variable) is an important technique for improving scientific accuracy. In *Fish Farm* (and definitely in real life), the same experiment rarely produces exactly the same value twice. Your objective in presenting these multiple results is to inform the reader about them while emphasizing the major trend.

### Tables

If your number of replicates is small (one or two), one technique is shown in Table 7--list multiple results on one line of a table. For larger numbers of observations per treatment, another approach is to list averages and ranges (the lowest and highest values obtained at that value of the independent variable). For example:

**Table 8.** Yields (kg/hectare) in pond experiments raising catfish at different stocking densities. All experiments used 30% protein feed, no groundwater input, and aeration when oxygen reached 5 mg/l. Four experiments were run at each stocking density.

| Fish per Hectare | Mean Yield | Yield Range |
|---|---|---|
| 2,500 | 1644 | 1422-1801 |
| 5,000 | 2892 | 2509-3790 |
| 10,000 | 5769 | 5110-6522 |
| 15,000 | 7183 | 6000-8999 |
| 20,000 | 8092 | 6035-10110 |
| 25,000 | 6740 | 2304-9089 |

Note that the Yield Range column adds valuable information. As the stocking density increases, the difference between the best experiments and the worst ones gets larger, probably because of disease outbreaks in the highest stocking densities.

## Graphs

Do *not* plot only a series of average values of the dependent variable. *All* data points should appear on a graph, and you should attempt to draw a smooth curve through the "cloud of points" (see Figs. 4 and 5). If the data are *wildly* scattered and the relationship appears to be so irregular that no reasonable smooth curve is possible, connect the average values of the points at each value of the independent variable:

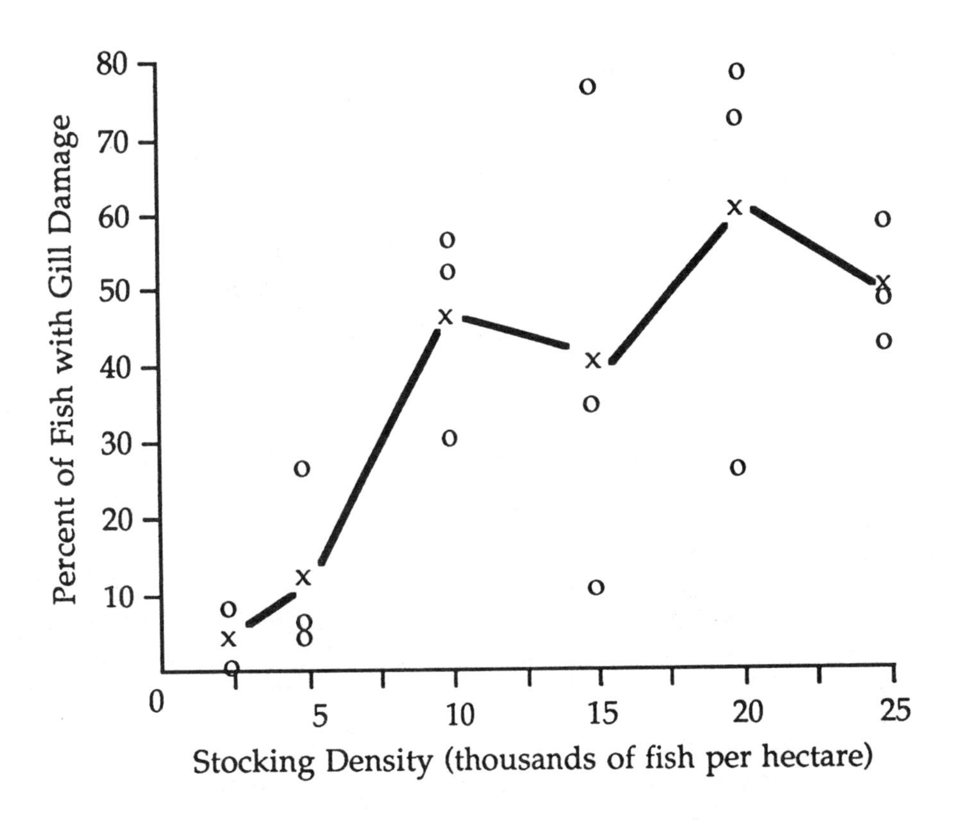


**Figure 11.** Percent of catfish exhibiting typical ammonia-induced gill damage at several stocking densities. o = an observation and x = the mean of the observations at that stocking density. There was no groundwater input, fish were fed 30% protein food, and aeration began if the dissolved oxygen reached 5 mg/l.

Again, no matter how variable the data are, never forget that your data presentation objective is to let the reader quickly see the underlying relationship between the dependent and independent variable.

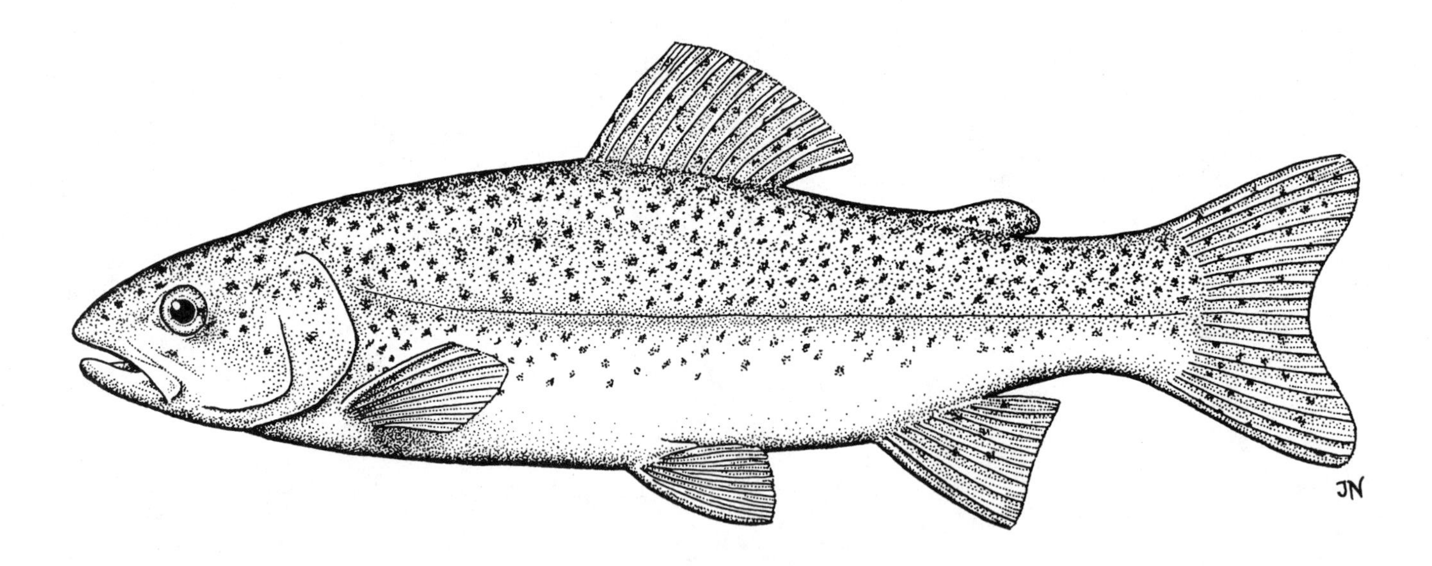


**Rainbow Trout** ***(Salmo gairdneri)***

The words *rainbow trout* conjure up the picture of a fly fisherman standing in a beautiful, remote stream, but 95% of all trout eaten in the U.S. are aquacultural products. Trout are disease-prone and require clean, cold, running water and high-protein feed, but their meat is so highly esteemed by consumers that culturists go to great expense to raise them. The center of trout production in the U.S. is the Snake River Valley of Idaho, where massive supplies of clean, chilled water gush out of the valley walls. Other production centers are in Utah, Pennsylvania, North Carolina, and Washington; trout are raised abroad in Norway, France, Spain, and Italy.

## Worksheet 2

Name ______________________________

1. Name one dependent, one independent, and one controlled variable in the **temperature** experiment performed by your group.

2. Did your temperature experiment have *more than one* dependent variable? Independent variable? Controlled variable? If so, name some of them.

3. Classify the following variables from the tank experiments as either continuous or discrete: type of fish, temperature, feed protein percentage, fish weight.

4. On a separate piece of graph paper, present the following data (labeled as Figure or Table 1): In the fall semester of 1992, the three most profitable teams using Fish A made $1.98 million, $1.92 million, and $1.87 million in their five-year production run. The top three teams using Fish B made $0.41 million, $0.37 million, and $0.32 million. The first, second and third place teams for Fish C made $1.01 million, $0.92 million, and $0.89 million. None of the Fish A teams lost money, but money was lost by 12% of the Fish B teams and 3% of the Fish C teams.

5. Decide on the best way to present your weight and cost per kg of weight gain data from your team's **feeding** experiment. Then construct a table or graph on the separate piece of graph paper mentioned above. Label it as Table (or Figure) 2, and give it an appropriate caption (for a figure) or title (for a table).

INSTRUCTOR CHAPTER 4

# Grading the *Fish Farm* Reports

The specific information necessary for grading the *Fish Farm* reports is contained in Chapter 5 of the student manual, and especially on pages 87-89. This chapter contains some more general guidance and warnings about trouble spots that have been encountered at Clemson.

The most general proposition of all is that errors in *Fish Farm* reports are like crimes: it's better to prevent them than punish them after they've happened. An instructor who patrols the class during the *Fish Farm* labs can detect and correct most data-recording errors, like not recording controlled variables or the unknown fish which was used. Likewise, the purpose of the data presentation exercise is to give the students some instructor-corrected practice at making graphs and tables and identifying independent, dependent and controlled variables. Of course, for this practice to be effective, the instructor must correct and return the worksheets before the reports are due. Many mistakes in writing the discussion can be avoided by emphasizing the grading criteria in Chapter 5 of the student manual.

But assuming that *Fish Farm* work is complete and a stack of reports now sits before you, the following may be helpful.

## The Raw Data Section

This section might reveal a large gap between instructor priorities and student priorities. Instructors would prefer that the raw data section should be complete, orderly (in the sense that each data point should be clearly associated with the methods which produced it), and show evidence of thought. In contrast, students will tend to fixate on cosmetic neatness and having the right answers. Thus, they may take down their data on scraps of paper, and later neatly transfer only the "good" data points to the data pages. The "good" data points are the correct ones (e.g., "Halfway through the stocking density experiment, we found that our fish needed groundwater after all, so we threw out all the mistaken experiments which didn't have groundwater"). This practice of editing data before it is recorded is a bad habit, so all data should be entered on the data pages.

## The Results Section

In most cases, the results section will need four line graphs (for the temperature, oxygen, feeding, and stocking density experiments), plus a table showing the results of the production run.

At Clemson, it is required that all *relevant* data be graphed or presented in a table, but this might not include all data recorded. For example, if the students above did discover belatedly that their fish needed groundwater input, then they would be required to graph the temperature experiment data which led them to this conclusion. But they would not be required to graph *stocking density* experiments which did not include groundwater because these experiments had no role in determining optimum stocking density.

One common student graphing mistake is use of unequal intervals on the axes of graphs; another is obsessive connection of points, even points which are replicates. Both mistakes are illustrated below:

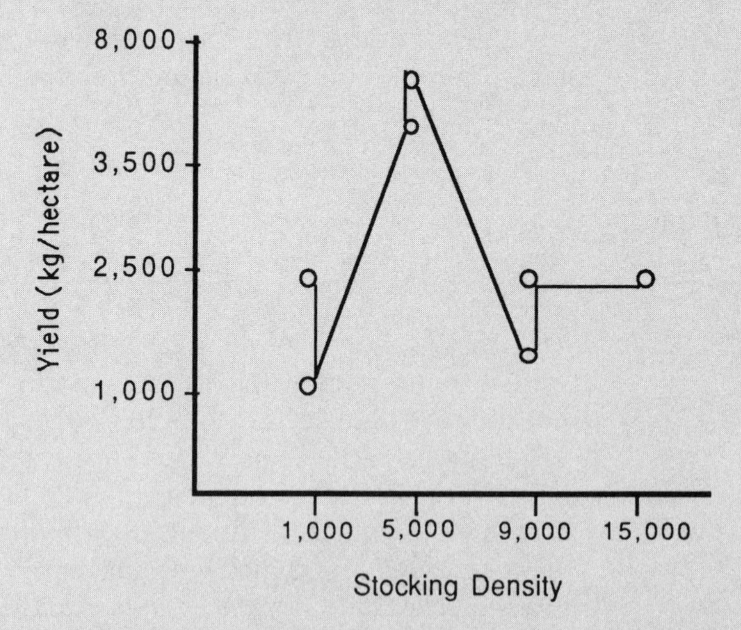


Both of these errors may reveal basic student misunderstandings. Unequal intervals may be used because the students perceive that the most important thing is to pair every x value with its corresponding y value, not to graph the points in correct relationship to each other. Connection of points may reveal an erroneous understanding of variability. Many students do not comprehend that while there may be a "true" mean profit or yield or cost per kg of weight gain for a given set of conditions, any experiment we perform can only estimate it with an observation which contains error. They tend to regard *Fish Farm* (or any scientific experiment) as a calculator which will give the uncorrupted right answer if the user pushes the right buttons. Therefore, they don't understand why they should do replicate experiments, they tend to be upset when the program reports two different outcomes for experiments done under the same conditions, they connect points rather than drawing a smooth curve through the cloud of points, and in tables they present a list of points rather than summarizing by presenting averages. Dealing with variability is a basic scientific skill, and so if *Fish Farm* brings these problems to light, your response can be a valuable teaching opportunity.

Other more minor errors in the results section are to give tables and graphs a short title rather than an informative caption, to fail to cite the conditions of the experiment in a caption which does exist, and to leave tables and graphs unnumbered.

According to the directions in the student manual, students do *not* have to write a results section narrative which summarizes the results. That function is accomplished in the discussion section.

## The Discussion/Conclusions Section

This section gives students the most trouble because, as many students have complained, "I just didn't know what to say." This complaint's credibility is reduced by the Grading Summary on pages 88-89 of the student manual. The Grading Summary lists the properties of a good discussion.

There are several common discussion problems. Students often do not make specific reference to their tables and figures (or only refer to them by saying something similar to, "My graphs show that my fish need..."). Many fail to discuss ambiguity and gaps in the data, perhaps because they don't understand that the result of an experiment is not a digitally correct answer but a suggestion or indication of the correct answer. If there is not enough replication, or if variability is large, or if data points are widely spaced, the suggestion is vague and unreliable. Finally, even when results of the most profitable teams are posted, many students do not compare their results with those of the "winners," and others merely state that the "winners" made so much while their team made somewhat less. At times the most profitable teams will use exactly the same conditions as less profitable teams, but most of the time the difference in profit can be traced to a difference in culturing conditions. The students should recognize this and point it out.

The conclusions section is often omitted by students, but if the report were really going to the management of an aquaculture company, the conclusions section would be vital.

At Clemson numerous references to the *Fish Farm* student manual are seen in reports, but there are few other citations. This is not a serious defect, since treatises on real-world aquaculture tend to be high-level and not directly relevant to *Fish Farm.* The correct listing of the manual in the Literature Cited section is:

Kosinski, R.J. 1993. *Fish Farm: A Simulation of Commercial Aquaculture.* The Benjamin/Cummings Publishing Company, Redwood City, CA.

Good luck!

CHAPTER 4

# The Pond Experiments

In the second *Fish Farm* lab, you will determine the correct number of fish to stock per hectare, and then you will test your previous work by determining the profitability of five years of simulated commercial operation (the production run) at your predicted optimum culturing conditions.

## Stocking Density Experiment

Stocking density (the number of fish which the culturist puts in per hectare of pond) is the most difficult culturing condition to determine correctly. The tanks you have used up to now had carefully controlled conditions, but in outdoor ponds, temperature, oxygen, ammonia and bacterial populations may easily attain levels which will cause massive fish kills. Of course, you will have tools to help you cope with these problems. You may vary the number of fish you put in the ponds, aerate the ponds if oxygen gets low, flush away wastes with rapid groundwater input, give the fish feed medicated with antibiotic if they get sick, and harvest either at the end of the growing season or when the fish reach marketable weight. But these tools must be used wisely if you want to make a profit.

When you deal with stocking density, you will be facing a dilemma: The more fish per hectare, the more product you will have to sell. But the more fish per hectare, the more poorly the fish will grow and the greater the chance for catastrophe due to oxygen depletion, fouling of the water with ammonia, epidemics, and resultant massive fish kills.

### Costs and Profits

To make a profit, a fish farm must sell its fish for an amount greater than the costs of producing them. Maximizing production is one part of the solution to this problem; the other is minimizing costs. Costs are of two basic types. First, **vari-**

**able costs** depend on the volume of fish raised, and would be zero if the farm lay idle. The cost of fingerlings and the cost of feed are the two variable costs which usually are 60% of the whole enterprise budget. **Fixed costs** (like property taxes, interest payments on loans, and costs of fixing leaks in the ponds) must be paid even if no fish are produced.

To make a profit the culturist must have sufficient product to cover both fixed and variable costs. *Fish Farm* will automatically compute profits as it simulates. Costs will start at a rather high value (representing fixed costs and the costs of the fingerlings) and will rise day by day (because of the added costs of feed). This means that profits will start at a negative value and will decline for the first several weeks. But then profits will begin to rise because the value of the stock is rising faster than costs. This happens because the fish are growing larger and so the potential harvest size is increasing. Second, the value **per kilogram** of the fish increases until the fish reach their marketable weight (454 gram or one pound for catfish). After reaching marketable weight, the value per kilogram remains constant. So there is a strong economic incentive to make sure that the fish reach marketable weight.

As an example of what you may expect, the profit picture of a typical catfish stocking density run is shown in Figure 12. Note how profit starts at a very low value because startup costs must be paid long before the fish are marketable, and how profits decline until mid-May. In mid-October the value of the stock exceeds expenditures for the first time, and at harvest profit stands at $212,349 from 200 hectares.

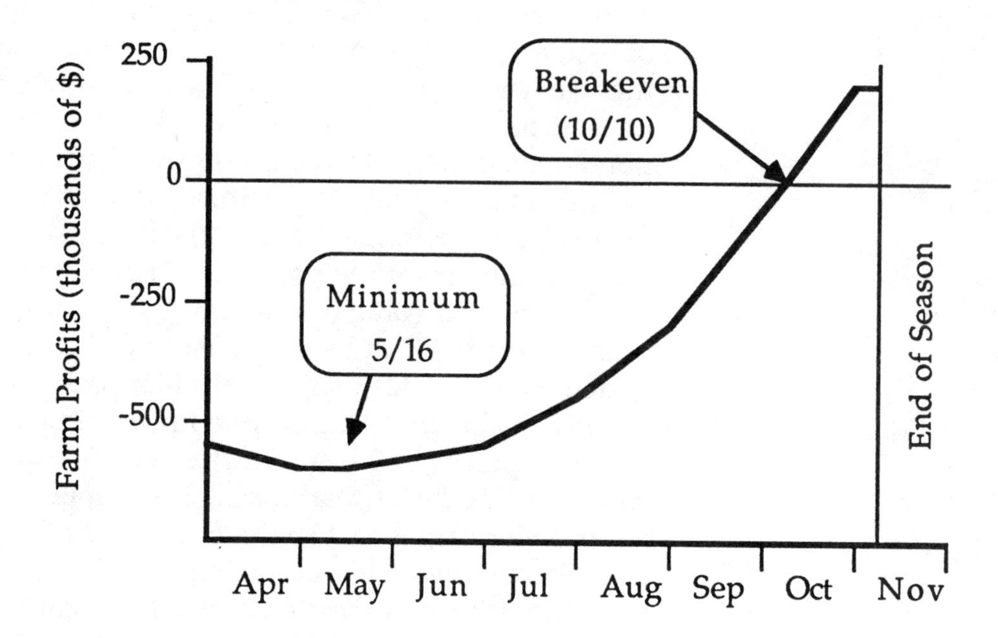


**Figure 12.** Profits for a catfish stocking density experiment. Catfish were stocked at 9,000 fish/ hectare, were fed 30% protein food at a varying rate, there was no groundwater input, and emergency aeration was applied if oxygen reached 5 mg/l. Harvest was on November 5.

## Time of Harvest

The fact that price per kilogram remains constant once marketable weight is reached causes a problem for some fish. While most fish show the profit curve in Figure 12, fish whose market price is low and who are inefficient feed converters have a profit curve more like that in Figure 13. They show peak profit not at the end of the growing season, but on the day they reach marketable weight, as shown in the circled section in the upper right hand corner of Figure 13.

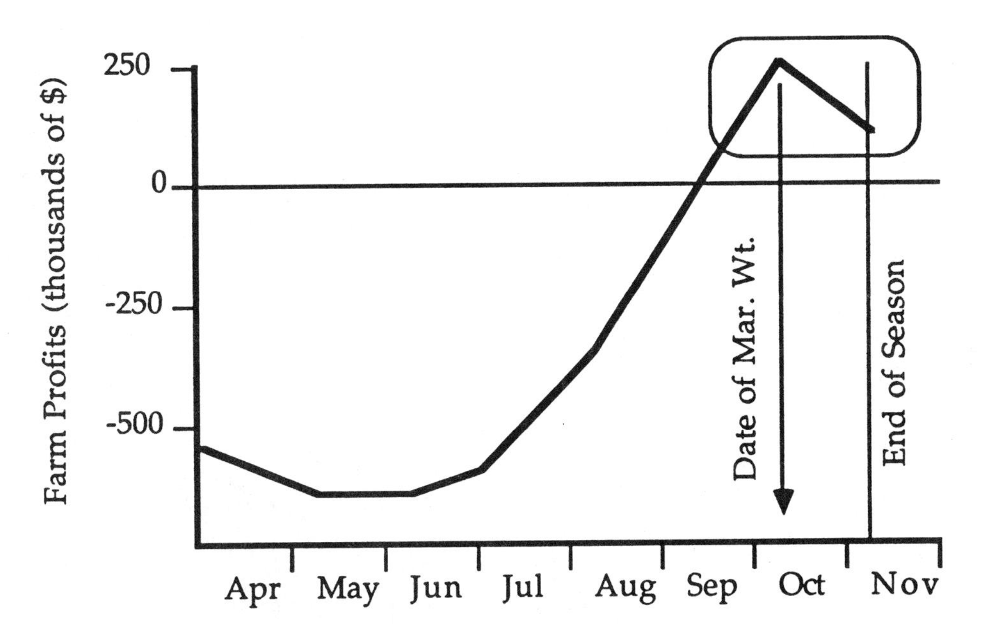


**Figure 13.** Profits for a stocking density experiment on an unknown fish which had a low feed conversion efficiency and a low price per kg. Peak profits occurred on the day the fish reached marketable weight.

Note that net profit declined from $252,000 on the day the fish reached marketable weight (500 g in this case) to about $90,000 on November 5, when they weighed 553 g. Obviously, this fish should be harvested on the day it attained marketable weight, not November 5.

The strange result that profit declines as the fish grow larger comes mainly from the fact that when fish are below marketable weight, their increase in value results from both an increase in weight and an increase in price **per kg** as they get bigger. Once they reach marketable weight, price per kg remains constant, so their increase in value is not as rapid. A decline in profits once marketable weight is exceeded is most likely to happen in a fish whose price per kg is not very much greater than the cost of the feed needed to produce a kg of weight gain. In a fish whose market price far exceeds the cost of its feed, profits will keep going up after marketable weight is attained, but at a slower pace.

If you find that your fish has a profit curve like Figure 12, you can instruct *Fish Farm* to harvest automatically on November 5. If you find the profit curve is more like Figure 13, you can instruct it to harvest automatically on the date that marketable weight is achieved.

## Variability in Stocking Density Experiments

Especially at high stocking densities, profit variability between runs can be extreme and cause students to have problems interpreting their data. Figure 14 shows profit results from 160 simulations, 10 at each of 16 stocking densities. Catfish were fed 30% protein feed at a varying rate. Aeration was begun if the dissolved oxygen fell to 5 mg/l. No antibiotics were used. Average profits and the range (difference between the maximum and minimum profits) are shown. Typical results from six *single* simulations from 5,000-10,000 fish per hectare are also shown.

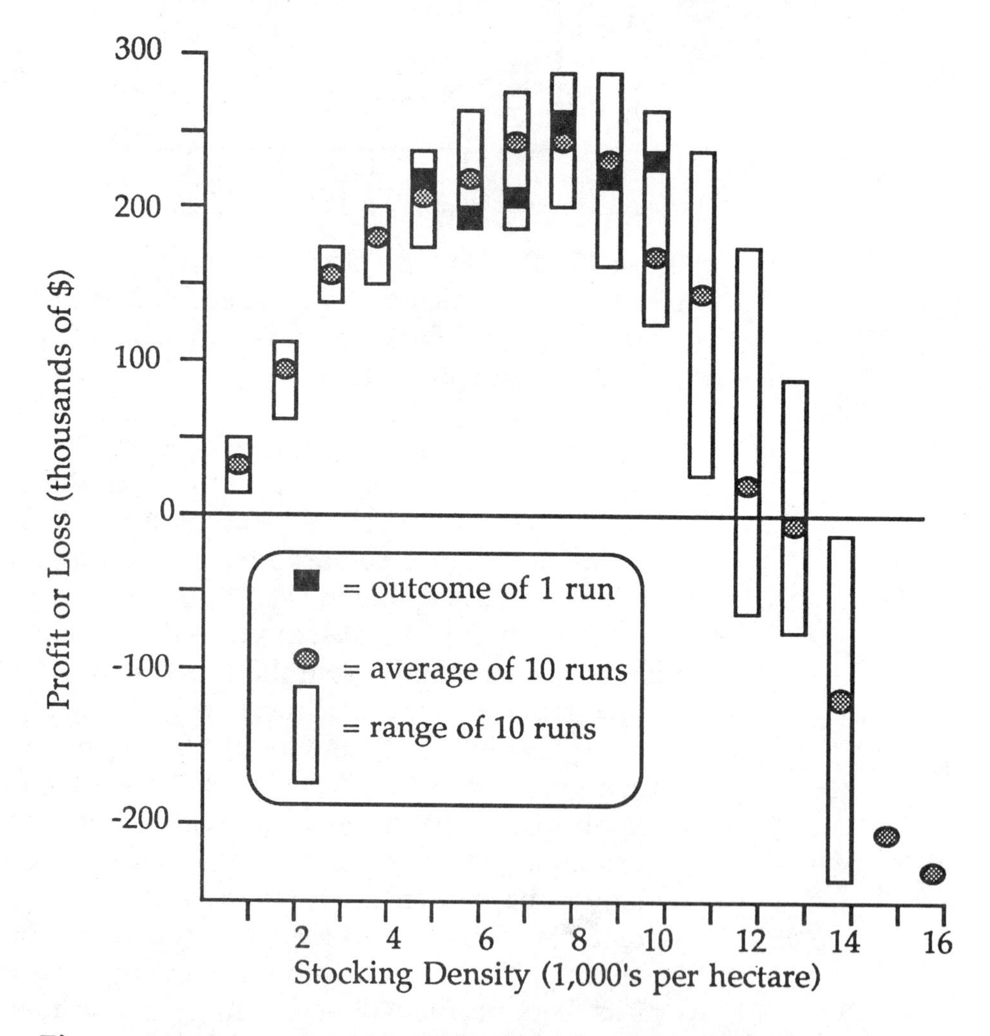


**Figure 14.** Average and range of profits from catfish stocking density experiments, computed from 10 replicated experiments per stocking density. Dark squares are results of typical single stocking density experiments.

The most obvious pattern in Figure 14 is that profit peaks at 7,000-8,000 fingerlings per hectare and declines rapidly once ponds are overstocked. But also note that variability between runs is large, and gets markedly larger at stocking densities which overcrowd and stress the fish. This occurs because each new batch of *Fish Farm* fish has a slightly different feed conversion efficiency and a different resistance to stress. Ranges are not shown for 15,000 and 16,000 fish per hectare because they would extend far off the bottom of the graph.

### Errors Caused by an Insufficient Range of Stocking Densities

The dark squares in Figure 14 show why students are sometimes baffled by their initial stocking density results. These squares show a typical student series of six single experiments from 5,000-10,000 fish per hectare. While on Figure 14 it is obvious that they are part of a larger pattern, Figure 15 shows the confusing picture which *Fish Farm* would show if asked to graph only these six points:

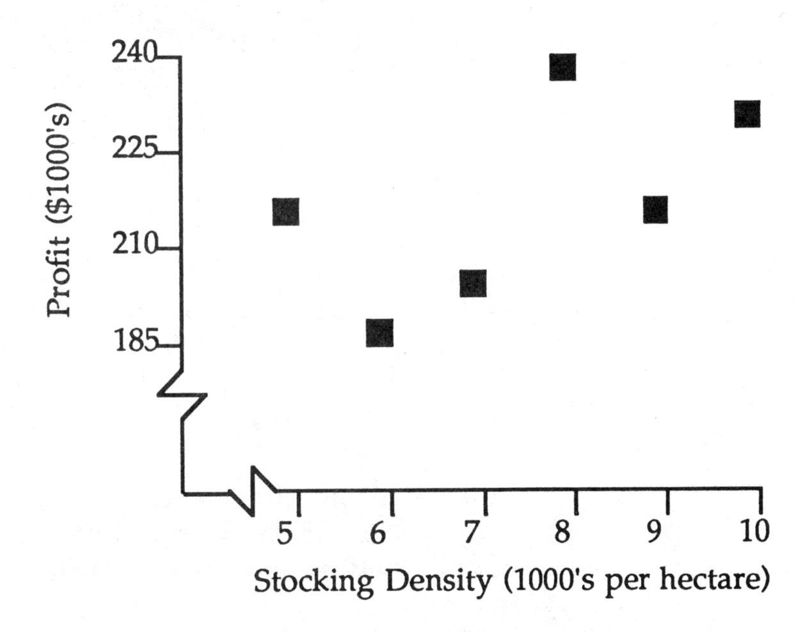


**Figure 15.** Profit results of a typical student series of catfish stocking density experiments as they would be graphed by *Fish Farm*. This experiment includes only stocking densities close to the optimum (approximately 7,500 fish per hectare).

Although the range chosen by the students is a good one (profit peaks right in the middle of it), there seems to be no pattern because of variability around the average and the fact that profit is uniformly high in this range of stocking densities.

If the students had chosen a pair of extreme stocking densities (say 1,000 and 13,000 fish per hectare) in addition to these six points, they would have seen the much more informative picture in Figure 16:

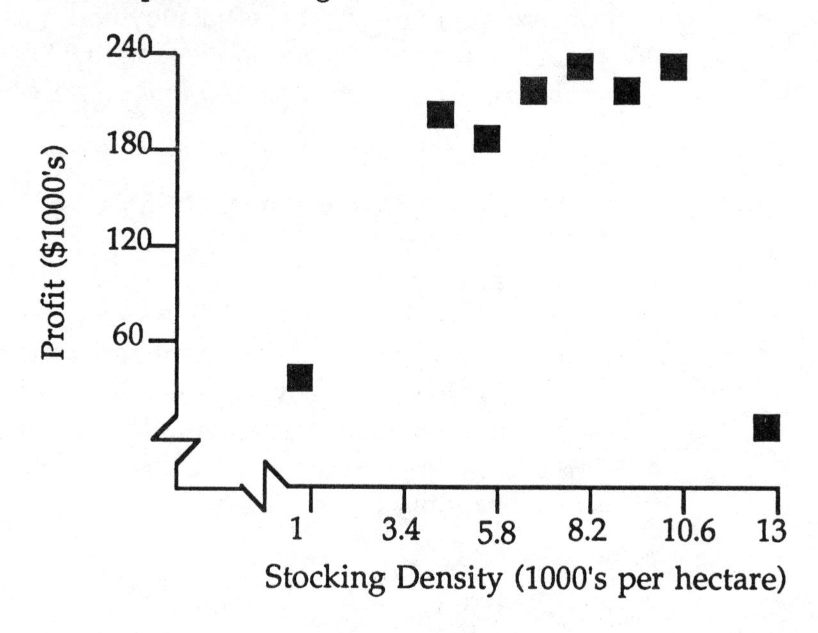


**Figure 16.** Set of student catfish stocking density experiments with inclusion of stocking densities both close to and far from the optimum.

In Figure 16, the level of detail is not great, but at least it is obvious that there *is* an optimum stocking density, and it lies somewhere between 5,000 and 10,000 fish per hectare.

Two morals:

1. Do several experiments at a stocking density. Make sure that a set of conditions produces reliable results before literally betting the farm on them.

2. To find the optimum stocking density, you must try stocking densities which are far from the optimum.

## Stocking Density and Groundwater Input

One way to overcome the problems brought on by high stocking densities is growing the fish in ponds or raceways with a fast rate of input of clean water. This allows a dramatic increase in stocking densities over those possible with static ponds. As well, a brisk groundwater flow through the pond will tend to moderate temperature fluctuations, bring the temperature towards the groundwater temperature (18° C in *Fish Farm*) and the oxygen toward the groundwater oxygen (9.5 mg/l in *Fish Farm*). Groundwater is most effective when it is used at the highest possible rate of flow (190% of the pond volume per day in *Fish Farm*).

But groundwater flow will greatly reduce the pond area. The 190%/day flow mentioned above will mean that the well will only be able to supply 1 hectare instead of the 200 hectares which would be available without groundwater input. Therefore, using groundwater input puts even more pressure on the farmer to maximize stocking density, and raceways generally tend to have stocking densities 10-100 times those of static ponds.

Static pond stocking densities which have been used in the recent past range from approximately 2,000 to 10,000 fingerlings/hectare for both carp and catfish, with a trend towards the use of even heavier stocking rates. In culturing arrangements with rapid water renewal, the rates are much higher. A high trout stocking rate might be 20 million fingerlings or 1 million adult fish/hectare of raceway, presuming about 1,600% water input per day (this water flow in *Fish Farm* would restrict the size of your "farm" to 0.1 hectares, the size of a small suburban lot).

Finally, water flowthrough will do nothing to reduce fighting and social stress. In fact, population densities usually have to be so high when rapid groundwater input is used that fighting will be worse than in a static pond.

## Possible Factors Limiting Profit

The growth rate of the fish (and profits) will always be limited by some factor. If this limitation is removed, another factor will soon become limiting. At times you will be able to do something about the limiting factor and at other times you will just have to accept its effects, but you should always be able to make an educated guess about what the limiting factor is.

### Numbers of Fish

If you stock very few fish, obviously profits will be limited. But in both the real world and in *Fish Farm,* overstocking is more common than understocking, so number of fish is rarely a factor limiting profit.

### Harvesting at the Wrong Time

Make sure that you harvest as profits peak, either at the end of the growing season or on the day marketable weight is attained. Look at the profit curve to determine whether your fish has a curve more like Figure 12 (harvest at end of season) or more like Figure 13 (harvest on day the fish reaches marketable weight).

### Food

In nature, lack of food is probably the most important factor limiting animal populations. Amount of food is usually not an important limiting factor in *Fish Farm* (you may supply all you want). However, the wrong **kind** of food (e.g., not enough or too much protein) may stress fish, cause lackluster growth and lead to disease outbreaks. Thus the importance of an accurate determination of the correct protein percentage in the feed must be emphasized again. If the fish seem to be getting sick for no apparent reason, suspect a feed protein content problem.

### Oxygen

Lack of oxygen is the major limiting factor in aquaculture ponds. As the growing season progresses and more and more feed is poured into the ponds, bacteria and algae reach high populations. Due to respiration and algal blooms, oxygen concentrations drop and become more and more variable. By late summer and early fall, the low points in the oxygen fluctuations (following collapse of algal blooms) are killing the fish, or are weakening them so disease outbreaks occur. The more the pond is overstocked, the more severe these problems are.

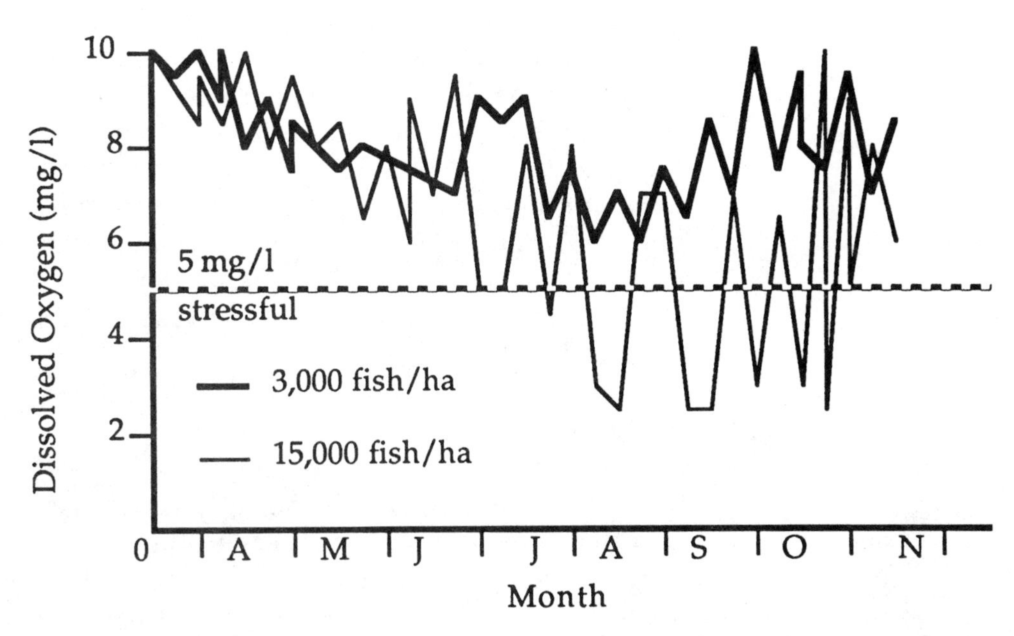


**Figure 17.** Dissolved oxygen concentrations in pond simulations containing 3,000 catfish per hectare (dark line) and 15,000 catfish per hectare (light line). **No emergency aeration was used in either case.** Catfish are stressed by dissolved oxygen concentrations below 5 mg/l. The 3,000 fish/hectare simulation had no mortality; the 15,000 fish/hectare simulation had 10.6% mortality, 57% of which was due to low oxygen.

Mechanical aeration is a partial solution to oxygen depletion problems, and was very important in raising catfish yields to their present high levels. But if the oxygen consumption rate in the pond is sufficiently great, the dissolved oxygen can be forced below the desired concentration even if the aerators are running night and day. And even with a very large aera-tor, other factors than oxygen will become limiting. If the 15,000 fish/hectare case above is run with emergency aera-tion, the mortality is still 1.5%, mostly due to disease trig-gered by high ammonia concentrations. The ultimate solu-tion is better water quality through reduced stocking density.

## Ammonia

When proteins are broken down by either fish or microorganisms, ammonia ($NH_3$) is released. Ammonia is very toxic to fish, causing reduced growth at low concentrations and gill damage, metabolic disturbance, and suppression of the immune system (leading to disease) at high concentrations. Fortunately, ammonia is constantly being removed from the pond by loss to the air and oxidation into harmless nitrate ions, but it can easily reach toxic concentrations if the input of protein in the feed is high. Another undesirable effect of ammonia is that its oxidation into nitrate consumes 3.8 grams of oxygen per gram of ammonia, explaining why oxygen problems are markedly worse when high-protein food is being given. Ammonia toxicity can never be eliminated, but to manage it you can reduce stocking densities, flush the pond with groundwater, and use the lowest possible protein percentage in your feed.

### Fighting and Social Stress

Some fish (such as catfish) placidly accept very high population densities and only hurt each other accidentally (as when they are pushed together and stab each other with their fin spines). But some species have a strong instinct to exclude all other fish from a home territory. If the population density is so high that only a few fish will be able to have a territory, the excluded "losers" will suffer stress and possible physical injury. Even the "winners" might spend most of their energy fighting and patrolling the boundaries of their territories. The result will be mortality, slow growth, and disease outbreaks due to a stress-related suppression of the fish immune system. And once fighting breaks out, it usually gets worse as the fish grow because larger fish demand larger territories, and bigger fish are more dangerous to each other.

The only way to banish fighting is to reduce the stocking density.

### Disease

The most common cause of massive mortality in *Fish Farm* is disease outbreak. Disease is very dangerous because it feeds on itself--the sicker the fish are, the less they are able to resist disease, and mortality rapidly accelerates into a catastrophic fish kill. In *Fish Farm*, disease is most serious when both fish stress (especially from low oxygen and high ammonia) and the population of bacteria in the water are high. Furthermore, each batch of fish has a different susceptibility to disease. Therefore one experiment using a heavy stocking rate could produce a good yield while another one with the same stocking rate could end in disaster.

If disease strikes, first ask yourself if it is serious (most disease outbreaks never become epidemics). If the program is telling you that two or three fish are dying per day and you have stocked 20,000, don't worry. You could take these losses for 100 days before even 1% of the fish had died.

If disease losses are serious, you have two options. First, in the long run, you could reduce the feeding rate in order to reduce bacterial populations and stop oxygen depletion. With catfish, a switch to "bare-bones" rations--a constant rate of 1% of biomass per day--will greatly reduce pollution in an overstocked pond. It will also reduce the harvest, but it is the only true solution. Secondly, one of the pond menu options allows you to give the fish a 10-day treatment with feed which has been medicated with antibiotics (in catfish culture, terramycin is often added to feed). In *Fish Farm*, using the antibiotic will stop the epidemic, at least temporarily. The disadvantage is that using medicated feed increases feed costs by $0.35 per kg, more than doubling the cost of typical feed. Because it is almost impossible to make a profit if you feed antibiotics for long periods, you should not use antibiotics more than once or twice during a growing season.

The program also will not allow you to use antibiotics in the last month of the run because then your fish would contain antibiotic when they were sold. Then the bacteria resident in the humans who consumed the fish might become resistant to the antibiotic, with resultant danger to the health of your customers.

## Troubleshooting Poor Profits

You should be able to earn profits of about **$250,000** per growing season with each of the *Fish Farm* unknown fish. If you have difficulty reaching this goal, see if your situation matches one of the following:

**Fish reach marketable weight by the end of the growing season and there is little mortality, but there are no profits.**

This could be one of two problems. First, the stocking density might be too low. Second, despite a correct stocking density, production costs might be too high. The most common reason for excessive costs is using a protein percentage which is either too high (making the food too expensive) or too low (worsening the FCR because the fish have to eat more in order to get proper nutrition). A less common reason for high costs is excessive use of antibiotics to combat disease.

**Fish fail to reach marketable weight and may suffer mortality.**

All the unknown fish should reach marketable weight in the 225 day *Fish Farm* growing season. If they don't, this will have a severe impact on profits. The simplest reason for slow growth is unsuitable temperatures, as when a warmwater fish is erroneously cultured with rapid groundwater input, or when a coldwater fish without groundwater input stops growing during the hot part of the year. But the most common reason for failure to grow is ammonia poisoning. Ammonia poisoning occurs when the fish are overstocked, and is worsened when the protein percentage is too high. Fighting between the fish (also due to overstocking) causes slight mortality and slow growth.

**Fish suffer heavy mortality mortality.**

To help diagnose the causes of mortality, *Fish Farm* provides a "Patholgist's Report" screen which indicates the percentage of fish death which can be ascribed to various causes (disease, low oxygen, high ammonia, nutritional deficiency, fighting, cold stress, heat stress, or rapid temperature change). Disease is the most common cause of death, but disease in *Fish Farm* is always triggered by other stresses. In most cases, these other stresses will be low oxygen and ammonia toxicity. The solution for the oxygen problem is mainly a lower stocking density, but at times setting the aerator to turn on at a higher dissolved oxygen concentration will help. Ammonia toxicity can also be reduced by a lower stocking density and by feeding at a lower rate. Sometimes the report will state that nutritional deficiency was a contributing cause to disease, but this usually means that the fish went off their feed due to low oxygen, high ammonia, or other bad environmental conditions.

## Arbitrary Shifts in Protein Content

In an ideal world, you would have accurately determined the optimal groundwater input, aeration level, and feed protein percentage before beginning the stocking density experiment. But sometimes the estimate of the optimum protein percentage is not correct, even if the protein experiment is performed correctly. Since using an incorrect feed protein percentage is a serious problem in the stocking density experiment, you should do some **arbitrary shift** experiments to verify that you are using the right protein percentage.

To do these experiments, find a rough stocking density which maximizes profits, and then keep the stocking density the same and shift feed protein content 3-4% higher and 3-4% lower than the value you had been using. If one of these shifts produces a marked increase in profits, shift protein an additional 3-4% in the direction which produced the increase, and keep repeating this process until a shift produces a reduction in profits. Discard your old protein percentage and use this most profitable percentage for all subsequent work. If, on the other hand, all protein shifts only produce a reduction in profits, your initial protein percentage was probably optimum.

One warning: If you *do* find that you need to change your protein percentage, clear your pond database and start the stocking density experiments over again. Once you change any condition (like protein), the new data points cannot be graphed along with the old ones, and the new optimum stocking density will probably differ from the rough optimum density you were using before.

## The Production Run

The production run will be five successive pond simulations (just like the stocking density simulations) which will extend from March 21 until November 5. All of them should be performed at what you predict are the economically optimum culturing conditions, based on your previous data. You will total these five years of profits and report them to your instructor, who will compare your total with the profits reported by other teams who were working on your fish. If you have experimented properly, the production run should report excellent profits.

But we must emphasize that **you will not be graded on profits**. You will be graded on the quality of your experimental planning, data evaluation, and logic, as described in your *Fish Farm* report. In other words, the production run **should** be profitable, but even if it ends in bankruptcy you can still get a good grade by giving your instructor a knowledgeable, logical account of **why** it ended in bankruptcy. Thus during the production run you will need to take more notes, rely more on the "Data Review" feature (present on the IBM and Macintosh versions of the program), and carefully note the "Pathologist's Report" information on the causes of fish death.

## Worksheet 3

Name ______________________________

Answer the following questions and be prepared to give this completed worksheet to your instructor **at the beginning of** the stocking density/production run session with *Fish Farm.*

1. Why will your profits suffer if you stock the fish at too **low** a density? At too **high** a density?

2. What factors tend to cause outbreaks of disease? What can you do about disease?

3. Why does groundwater input allow far higher stocking densities than can be attained with static ponds?

4. What are the arbitrary shift experiments which you do in the middle of the stocking density exercise? What results of these experiments would convince you that you need to change your protein percentage?

5. What are some typical stocking densities (fingerlings per hectare) that have been used with both static ponds and raceways?

## Stocking Density Experiment Procedure

> **Your Stocking Density Objectives:**
>
> Determine whether your fish should be harvested when they reach marketable weight or on November 5.
>
> Find the stocking density which gives your farm the highest *reliable* profits.

1. If your computer is off when you start, skip to step 2. If your computer is on and still running *Fish Farm* with the fish your instructor assigned to your team, indicate that you wish to perform a **pond experiment**. Then skip to step 7.

2. If your computer is off when you start, place your *Fish Farm* disk in the disk drive of the computer and turn the computer and monitor on.

3. After a title screen, you may be asked to adjust the program speed so a second timer ticks at the correct rate. Follow the directions on the screen.

4. **Apple II users only:** Indicate that you wish to work with **outdoor ponds**. This question will come later for IBM and Macintosh users.

5. Select the fish which your instructor assigned to your team. Read the description for hints on the culturing conditions the fish will require. Then accept the fish.

6. **IBM and Macintosh users:** Indicate that you wish to work with **outdoor ponds** after your choice of fish.

7. The program will then display the main pond menu. Decide on at least five stocking densities (fish/hectare) with which you wish to start. The densities should range from understocking to overstocking so you can present a convincing case that you have found the

optimum density. List these five stocking densities on pages 67-68, the stocking density data pages. Also fill in the percent protein in the feed and the percentage input of groundwater for the first density. Then choose **stocking density** and enter your first density into the computer.

## Stocking Density Data for Fish _____

Controlled Variables: Aeration Oxygen __________

Harvest (ES or MW) __________

| Fish/Hectare | Protein | Grdwtr % | Average Wt | FCR | Mortality | Profit |
|---|---|---|---|---|---|---|
| | | | | | | |

Comments: ______________________________

| | | | | | | |
|---|---|---|---|---|---|---|
| | | | | | | |

Comments: ______________________________

| | | | | | | |
|---|---|---|---|---|---|---|
| | | | | | | |

Comments: ______________________________

| | | | | | | |
|---|---|---|---|---|---|---|
| | | | | | | |

Comments: ______________________________

| | | | | | | |
|---|---|---|---|---|---|---|
| | | | | | | |

Comments: ______________________________

| | | | | | | |
|---|---|---|---|---|---|---|
| | | | | | | |

Comments: ______________________________

| | | | | | | |
|---|---|---|---|---|---|---|
| | | | | | | |

Comments: ______________________________

| | | | | | | |
|---|---|---|---|---|---|---|
| | | | | | | |

Comments: ______________________________

| Fish/Hectare | Protein | Grdwtr % | Average Wt | FCR | Mortality | Profit |
|---|---|---|---|---|---|---|
| | | | | | | |
| | Comments: | | | | | |
| | | | | | | |
| | Comments: | | | | | |
| | | | | | | |
| | Comments: | | | | | |
| | | | | | | |
| | Comments: | | | | | |
| | | | | | | |
| | Comments: | | | | | |
| | | | | | | |
| | Comments: | | | | | |
| | | | | | | |
| | Comments: | | | | | |
| | | | | | | |
| | Comments: | | | | | |
| | | | | | | |
| | Comments: | | | | | |
| | | | | | | |
| | Comments: | | | | | |
| | | | | | | |
| | Comments: | | | | | |
| | | | | | | |
| | Comments: | | | | | |
| | | | | | | |
| | Comments: | | | | | |

8. Choose **feeding instructions** and enter the optimal percent protein which you determined during the feeding experiment. Indicate that you wish the feeding rate to **vary**.

9. If you wish to use any groundwater input, select the groundwater option and enter the percent of the pond volume which you wish to input per day.

10. Select the **emergency aeration** option. Review your data from the oxygen experiment and enter the oxygen concentration at which your fish begins to be seriously stressed. Also write this value on the top of the stocking density data sheet.

11. If the **time of harvest** is set on **End of Season,** leave it there for now. If it is set on **Marketable Weight,** change it to **End of Season**. You will use the first stocking density experiment to determine what the time of harvest should be. Leave **Harvest (ES or MW)** on the stocking density data sheet blank.

12. Press **ESC** and the simulation will begin. The data portion of the pond display screen appears below.

| **Catfish** | **3/24** |
|---|---|
| Water temperature (C): | 17.0 |
| Dissolved oxygen (mg/L): | 9.2 |
| Feeding (% of biomass): | 1.48 |
| Feed given (kg/hectare): | 1.6 |
| Fish weight (g): | 28 |
| Surviving fish per hectare: | 7500 |
| Biomass (kg/hectare): | 213 |
| Feed conversion ratio: | 1.1 |
| Profits from 200 hectares: | -466066 |

This screen was explained in the discussion of the tank experiments.

13. As the simulation proceeds, remember that you can pause it. Also, you can see graphs of the oxygen in the pond, the water temperature, the mean fish weight, and profits. Follow the directions on the screen to do this.

14. If fish start to die, a message telling the number of dead fish will be flashed on the screen. Whatever kind of stress is causing the deaths (e.g., disease, low oxygen, etc.) will also be flashed on the screen. If the number of dead fish seems large, you may pause the simulation and consider your options:

    a. If disease is serious, you may press the **ESC** key, return to the main pond menu, and give antibiotic in the feed. This may help in the short run, but the long-term solution is probably to reduce the stocking density.

    b. If disease is not a factor, the problem is probably low oxygen or ammonia poisoning, and the **only** solutions are to increase groundwater input or to reduce stocking densities. However, you may not change groundwater input in the middle of a run. If the fish are fighting, the only solution is to stock at a lower stocking density in the next experiment.

    c. If the experiment is a total disaster and fish are dying by the thousands, you may return to the pond menu and harvest the fish early.

15. Just before the fish reach their marketable weight, select the **profit graph**. Watch to see if profit peaks and then starts to decline. **Apple II users:** Noting this will be especially important for you.

16. When the simulation reaches November 5, it will stop automatically on the harvest screen. The harvest report will give you the history of the experiment, including the final weight of the fish, the FCR, the kg of fish harvested, the mortality rate, and the profit and loss figures. If fish died, a second screen will give a "Pathologist's Report" showing the causes of death. Record the final fish weight, the FCR, the mortality rate, and the profit

on your stocking density data sheet. Record losses as negative profits. Record the causes of mortality or any other notes on the **Comments** line for that stocking density.

**17.** **Add** the results to your pond database (if the experiment was a valid one). However, do not proceed to the next experiment yet.

**18.** You must determine time of harvest:

a. **Apple II users:** If the profit peaked as the fish attained marketable weight and then declined for the rest of the simulation, the fish should be harvested as they reach marketable weight. On the other hand, if the profit continued to rise after the fish reached marketable weight (which will be true in most cases), they should be harvested at the end of the season. **If the fish never reached marketable weight or only reached it at the very end of the experiment, no conclusions about time of harvest can be drawn.** Apple II users should skip to step 19.

b. **IBM and Macintosh users:** Press **R** to review the results of the experiment, and select the review of average fish weight.

c. **If the fish never reached marketable weight or only reached it at the very end of the experiment, no conclusions about time of harvest can be drawn.** In this case go on to step 19.

d. However, if marketable weight was reached before harvest, note the approximate date on which marketable weight was attained.

e. Then press **ESC** and request the profit graph. Determine if profit began to decline on the date marketable weight was reached. If it did, the time of harvest should be **Marketable Weight**. If the profit kept rising after marketable weight was attained, the time of harvest should be **End of Season**. But be careful--sometimes the requirement to harvest at marketable weight only becomes clear when the fish are stocked at close to their optimum density. Even a fish which requires marketable weight harvesting may increase in profits to the end of the growing season if it is stocked at below optimum density.

**19.** Exit from the data review menu, choose to "continue to the next experiment," indicate you wish to **perform another experiment**, and repeat steps 7 and then steps 12-17 with the other stocking densities you have chosen. There is no need to reenter the information on diet, aeration, etc. Keep in mind that if you changed the harvest time to "Marketable Weight," then the fish will be harvested whenever marketable weight is attained rather than on November 5.

**20.** After you have completed the five initial stocking densities, select "Review your past pond experiments" from the end-of-experiment menu. Then view the graph of profit vs. stocking density. If any of your experiments incurred large losses, these points may force all the profitable points into a tiny region at the top of the graph. In that case, you may wish to delete the "outliers" from your pond database. Follow the directions on the screen to do this. However, the outlier points are still part of your data, so don't delete them from your data pages.

**21.** Use the profit vs. stocking density graph to select additional stocking densities to test. Also, since experiments at the same stocking density can vary so much from one another, you should do several replicate experiments at stocking densities which seem to be the most profitable.

**22.** If you change a condition such as feed protein percentage or dissolved oxygen at which aeration begins, the stocking density experiments after this point cannot be compared with the ones before the change. **Clear your pond database before adding the new data to it.** You should graph only comparable data points together.

**23.** If you have determined a rough optimum stocking density, perform the experiments on **arbitrary shift** in feed protein content which are described on page 63. **If you end up deciding on a new feed protein**

**content, clear your pond database and start the stocking density experiments again.**

24. Before you move on to the production run, ask your instructor to look at your profit vs. stocking density results. You may also look at graphs of fish weight vs. stocking density and total harvest vs. stocking density.

You now have all the data you should need for culturing your fish profitably. Discuss your data with your laboratory instructor and with other student teams in the lab who are working on the same type of fish. Use these discussions to establish the conditions you will use for the production run, which will be done in the same laboratory period as the stocking density exercise.

## The Production Run Experiment Procedure

> **Your Production Run Objective:**
>
> Perform five simulations which earn the highest possible profits. All five runs need not be performed under the same conditions, but you may do only five consecutive runs.

1. If your computer is on and still running *Fish Farm* with the fish your instructor assigned to your team, skip to step 6. Otherwise, go to step 2.

2. If your computer is off when you start, place your *Fish Farm* disk in the disk drive of the computer and turn the computer and monitor on.

3. After a title screen, you may be asked to adjust the program speed so a second timer ticks at the correct rate. Follow the directions on the screen.

4. **Apple IIe and Apple IIGS users only:** Indicate that you wish to work with **outdoor ponds**. This ques-

tion will come later for IBM and Macintosh users.

5. Select the fish which your instructor assigned to your team.

6. **IBM and Macintosh users:** Indicate that you wish to work with **outdoor ponds** after your choice of fish.

7. The program will then display the main pond menu. Choose **stocking density** and enter your chosen optimal stocking density.

8. Choose **feeding instructions** and enter the optimal percent protein which you determined during the feeding exercise, perhaps as modified by the results of the arbitrary shift experiments. Indicate that you wish the feeding rate to vary.

9. If you wish to use any groundwater input, select the **groundwater** option and enter the percent of the pond volume which you wish to input per day.

10. Select **emergency aeration**. Review your data from the oxygen experiment and enter the oxygen concentration at which your fish begins to be seriously stressed.

11. Select **time of harvest**. For most fish, **End of Season** is the best harvest time. However, if profit for your fish starts to decline after marketable weight is reached, set time of harvest to **Marketable Weight**.

12. Note all these run conditions on the production run data table (page 75). Then press ESC and the simulation will begin.

13. During the simulation, if you change any of the pond conditions (such as the aerator setting), note this fact and the date. If you treat the fish with antibiotic, note the dates on which antibiotic use was begun. Also, if large numbers of fish start to die, note the date and the cause. You may pause the simulation to make these records.

## Production Run Data for Fish _____

**Controlled Variables:** Aeration Oxygen __________

Feed Protein __________

Harvest (ES or MW) __________

Groundwater % __________

| Fish/Hectare | Average Wt | FCR | Mortality | Profit |
|---|---|---|---|---|
| ________ | ______ | _____ | _____ | ________ |

**Comments:** ______________________________

| Fish/Hectare | Average Wt | FCR | Mortality | Profit |
|---|---|---|---|---|
| ________ | ______ | _____ | _____ | ________ |

**Comments:** ______________________________

| Fish/Hectare | Average Wt | FCR | Mortality | Profit |
|---|---|---|---|---|
| ________ | ______ | _____ | _____ | ________ |

**Comments:** ______________________________

| Fish/Hectare | Average Wt | FCR | Mortality | Profit |
|---|---|---|---|---|
| ________ | ______ | _____ | _____ | ________ |

**Comments:** ______________________________

| Fish/Hectare | Average Wt | FCR | Mortality | Profit |
|---|---|---|---|---|
| ________ | ______ | _____ | _____ | ________ |

**Comments:** ______________________________

14. The program will stop automatically on the harvest screen on either November 5 or the date marketable weight is achieved. The harvest report will give you the history of the experiment, including the final weight of the fish, the FCR, the mortality rate, the profit and loss figures, and the causes of death. Record all this on the production run data page, **not just on the production run data sheet** which your instructor will give you.

15. **Add** the results to your pond database if you wish. Then press the **spacebar** to continue to the next experiment.

16. Indicate that you wish to **perform another experiment**. If you wish to use the same stocking density and other conditions again, repeat step 7 with the same stocking density and then steps 12-15. There is no need to re-enter the information on diet, aeration, and so on if you do not change it. You **may** change any conditions you wish between runs, however. The only rule is that you may perform only five runs and you must use the profits (or losses) from all of them.

17. When you have finished all five runs, turn in a summary sheet to your instructor. Write the names of the team members, your lab section, the fish you were using, and then, for each of the five runs, the run conditions (feed protein percentage, groundwater input, aerator setting, time of harvest, and stocking density) and the run results (final fish weight, mortality rate, and the profit or loss).

Compare your results with other teams in the lab, and then discuss them with your laboratory instructor. The next chapter offers advice on the completion of the *Fish Farm* report.

CHAPTER 5

# The *Fish Farm* Report

Your work on *Fish Farm* will be graded by evaluation of a *Fish Farm* report. It should clearly and concisely summarize your culturing recommendations and the data which support them.

## Purpose of the Report

While aquaculture is an important and interesting subject, you did not do the *Fish Farm* exercises to learn how to culture fish. The purpose was to learn how to use scientific methods. Therefore, the highest grades will go to reports which describe an orderly series of experiments, use good data recording and presentation techniques, and reach conclusions which are clearly based on strong evidence.

In addition, to solve any problem efficiently, you must be familiar with the subject matter. Therefore another question involved in evaluation of your report is whether you can apply your knowledge of aquaculture to the interpretation of your data.

To summarize, your report should show logical, knowledgeable interpretation of your data, presented in a standard scientific writing style.

### The Report's Audience and Objective

Before beginning any piece of writing, you should determine the audience which you are attempting to persuade, inform or entertain. In *Fish Farm*, write as if you were informing the executives of the aquaculture company about the results of your experiments. You wish to clearly present the fish's optimum culturing conditions and the evidence on which you are basing your culturing recommendations.

### The Style of the Report

The major elements of the scientific writing style are clarity and conciseness. As one biology journal recommended, "Make sure that each sentence conveys the exact truth as simply as possible." Distinctive personal writing style, building suspense, and entertaining the reader are excellent ingredients in many kinds of writing, but please keep them out of your report.

## Contents of the Report

The report will have four parts: a cover page, the raw data as recorded on the temperature, oxygen, protein, stocking density, and production run data pages (30%), a Results section of summary graphs and tables on separate sheets of graph paper (30%), and one or two double-spaced, typed or computer-printed pages which discuss the data and state your conclusions (40%).

### The Cover Page

The cover page should be done in the format below. It is worth 2% of your grade. Use the name of your fish, not "catfish." And yes, it should be typed or printed.

```
Optimum Culturing Conditions

for

Catfish




by

Ima Student

Biology 101, Section 1

September 22, 1992
```

### The Raw Data Section

The data pages should record your experimental methods and results. Your record here need not be neat, but it must give the reader a legible account of which experiments were performed and what the outcomes were. Someone else who had access to *Fish Farm* and knew how to use it should be able to reproduce your results by consulting your data pages. Include every experimental condition you used for a series of experiments, at least at the beginning of your notes on those experiments. For example, even though you may be doing a temperature experiment, you must record the feeding conditions and the oxygen concentration. And don't make the most embarrassing mistake of all by never mentioning which fish you were using!

Part of an adequate set of notes about a series of protein experiments appears on the next page. Note that the entries are preceded by a very abbreviated description of the methods. Without this, all the data would be useless. Also note that the data entries are written down in the order in which they were taken, and are interspersed with comments and questions. This is perfectly acceptable (and even desirable) for the raw data section. Crossing out mistakes and moderate "messiness" are also acceptable. A data page should not be just a blind recording of data--it should show evidence of thought. And the thoughts need not always be correct. In the example below, the student is bewildered by the fact that the fish weight is the same but cost is going up as protein content changes. This is because the weight gain is due to food with progressively higher costs. The student doesn't seem to understand this now, but hopefully will by the time the Discussion is written.

## Feeding Data for Fish Catfish

**Controlled Variables:**

Temperature 25°
Oxygen 10 mg/l
Feeding Rate 2%/day
Constant or Varying? varying

| Protein | Weight | FCR | Cost of Weight Gain | Comments |
|---|---|---|---|---|
| 0% | 20 g | - | infinite | 7 fish died |
| 10% | 42 g | 3.6 | $0.90 | |
| 20% | 54 g | 2.5 | $0.60 | |
| ? 30% | 60 g | 1.4 | $0.42 | ASK TA |
| 40% | 62 g | 1.4 | $0.62 | |
| | Weight is more but cost is higher??? | | | |
| 50% | 62 g again? | 1.4 | $0.70 | |
| 60% | 62 g again??? | 1.4 | $0.90 | Why is cost going up if weight is same?? |
| ? 30% | 64 g | 1.3 | $0.40 | |
| ~~25%~~ | ~~4? g~~ | | | mistake |
| | | | | |
| | | | | |
| | | | | |
| | | | | |
| | | | | |

What is *not* acceptable in the raw data section is a record which omits data which is presented later in the Results section, or which so disorganized that it cannot be followed, or which has results which are not clearly linked with some record of the methods which produced them. A disembodied "56 g," by itself, has no value.

**Raw Data Points**

Of the 30 percentage points for the raw data section, 2% is assigned to the cover, 18% for completeness of data, and 10% for clear association of all data with the methods which produced them. You may also win up to 4% as bonus points if your notes show evidence of thought about the data (perceptive comments, questions, and so on).

**The Results Section**

If the raw data section was your working record of your experiments, the Results section is the formal, orderly record which you will present to other readers. It is a series of appropriately labelled tables and graphs, done in the format recommended in Chapter 4, and presented on sheets of graph paper which will be stapled to the Discussion section.

For example, after a few more points were added, the feed protein experiment presented above might be presented as a graph with the following format:

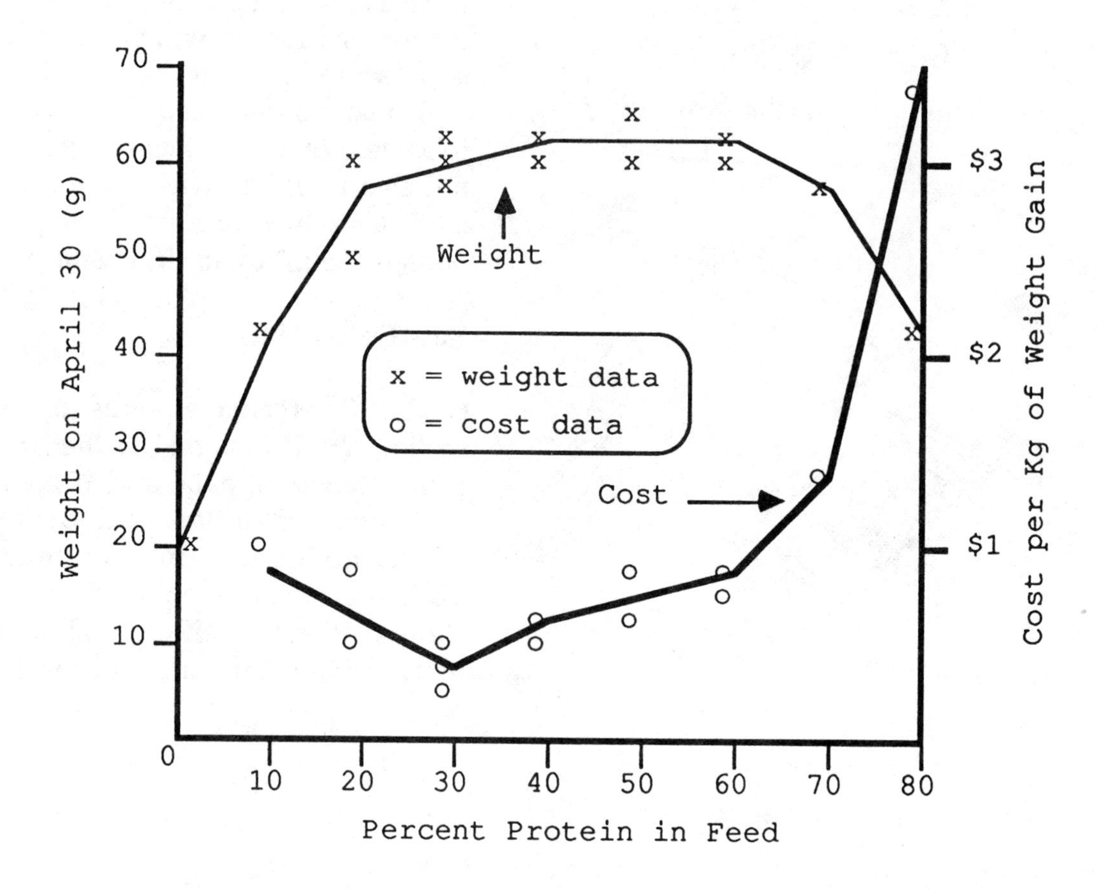


Figure 18. Growth of catfish as a function of protein content in the diet. Fish were grown in tanks at 25° C, 7 mg/l oxygen, and with varying feeding rate. o = weight observations and x = cost observations.

The weight curve is not necessary, since your protein decisions were reached on the basis of cost of weight gain only. However, it is an informative addition.

The graphs and tables in the Results section are meant to be presented to other readers, so they should should be neatly done, with lines drawn with a ruler, accurate scaling of axes and placement of points, and all labels clear and consistent. Also, most of the graphs which you do for your report will be line graphs showing how some variable (e.g., fish weight) responds to another variable (e.g., feed protein content). Here, you wish to emphasize the general trend, not the random variation around the trend, so you should draw a smooth curve through the cloud of points, not connect the points exactly.

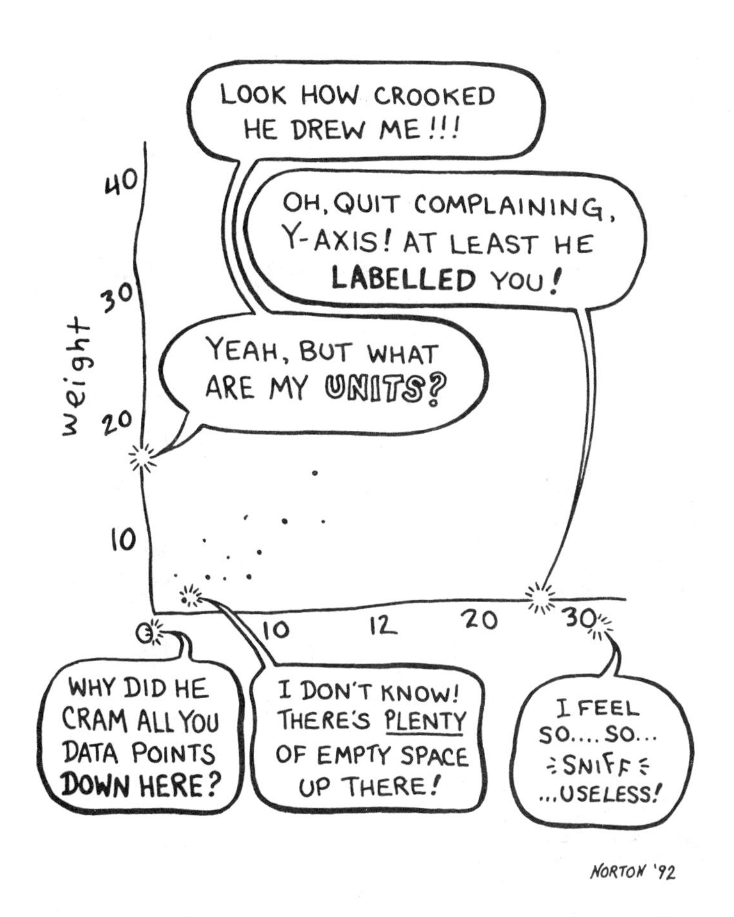


The data should be summarized and organized (e.g., summed, averaged, ordered by size of the independent variable, and so on) as appropriate. The Results should not just be a neater version of the Raw Data pages. For example, if the data in Figure 18 were presented in a table in the Results section, the table would show the **average** cost for each of the protein contents, proceeding from the lowest protein content to the highest.

As indicated in the Raw Data section, the reader should be able to identify every data point with the methods which produced it. Make sure that table titles, figure captions, column headings, and labels on axes indicate methods precisely. Figure 18 gives examples of how this is done.

### Results Points

The 30 percentage points for the results section will be awarded by subject. Temperature, oxygen, and protein results will get 5% each. Stocking density will get 10%, and the production run results will be worth 5%.

You **must** present either a table or a graph to be eligible for **any** points in any of the subject matter areas of the Results section. For example, if you present neither a table nor a graph for protein, you will lose those 5 points no matter how perceptive or correct your comments about protein are in the Discussion. In scientific writing, the data are always given to allow the reader to draw his or her own conclusions.

### Discussion and Conclusions

In this section, you should interpret the results, explain their significance for your culturing recommendations, and give your conclusions. This should be done with reference to the tables and graphs of the Results section. The Discussion and Conclusions section should only be one or two double-spaced typed pages long, so it must also be fairly concise. For example, the section of a Temperature/Oxygen/Feeding Discussion which dealt with the protein data above might read:

> It is important to feed the lowest possible protein content in order to reduce both feed costs and pollution. Figure 3 shows the results of growing our fish at feed protein contents ranging from 0% to 80%. While the greatest weight gains were produced at protein contents from 40%-60%, 30% clearly produced the cheapest weight gain. Because cost is considered a more important consideration than weight gain in determining optimum feed protein (Kosinski, 1992), we are recommending 30% protein for the feed.

Background material appears in Chapter 2 (tank experiments), Chapter 4 (pond experiments), and in the Appendix. You should use both your results and your knowledge of aquaculture to arrive at your culturing recommendations:

> Although the greatest profit during our stocking density experiments was produced by stocking 10,000 fish/hectare (Fig. 4), sometimes this produced severe disease outbreak, despite antibiotic treatment. Kosinski (1993) suggested that antibiotic is not a substitute for a good water quality, and recommended a lower stocking density as a

> long term solution. Therefore we are recommending 7,500 fish/hectare as the highest stocking density which will reliably produce profits and avoid disease.

Another purpose of the Discussion is to explain weaknesses in your data and discuss the possible effects on your conclusions. For example:

> These results do not include any experiments using groundwater input. This probably would have allowed us to use stocking densities at least ten times higher than are possible with static ponds (Limburg, 1980).

Finally, to put your experiment in context, it should be compared to other similar experiments. In your case, the obvious candidates are the experiments of the most profitable teams for your fish, if that information is available to you. For example:

> While we made $1,295,300 in our production run, the first-place team for catfish (from section 22) made $1,560,200. Their conditions seem very similar to ours except that they used 10,500 fish/hectare instead of 10,000, as we did. With more stocking density experiments, we might have reached the same conclusions and had higher profits.

The Conclusions should be a short paragraph on the bottom of the Discussion and should summarize your major recommendations, with brief reference to the results. If your report were really going to the management of an aquaculture company, you could be sure that more executives would read the Conclusions than any other section, and most would read **only** the Conclusions. For example:

```
    In conclusion, in the tank experiments
we found that catfish should be cultured in
static ponds with no groundwater input,
that aeration should begin when dissolved
oxygen reaches 5 mg/l, and that 30% protein
in the feed is best.  In the pond experi-
ments, highest profits were attained with
10,000 fish/ hectare.  These conditions
allowed us to earn $1,295,000 during our
production run.
```

**Discussion/Conclusions Points**

The Discussion/Conclusion is worth 40 points (35 for the Discussion and 5 for the Conclusion). There are also 5 possible bonus points. The Discussion should emphasize how culturing recommendations were derived from the data, and not on extraneous factors like the profit or details of how *Fish Farm* operates. The good Discussion makes frequent reference to the tables and graphs of the Results. It should devote more attention to explaining especially critical or ambiguous results (e.g., a persistently unclear optimum protein percentage) than to harping on obvious results (e.g., temperature if it is clear that the fish does or does not need groundwater input). It should discuss possible weaknesses in the experiments, and compare the results to those of other student teams, if this is possible. Finally, it should show knowledge of aquaculture. Up to 5 bonus points are awarded to reports which cite relevant material which is not from the *Fish Farm* manual.

Good scientific writing style, lack of errors and neat appearance will earn you points for good presentation.

The 5 points for the Conclusions are awarded when the Conclusions section is present and its recommendations agree with those of the Discussion. Many students leave out the Conclusions and lose these 5 points. *Be sure to include a Conclusions section.*

The Conclusions may be followed by a Literature Cited section; see the literature citations at the conclusion of this chapter. The correct citation for this manual is:

Kosinski, R.J. 1993. *Fish Farm: A Simulation of Commercial Aquaculture.* The Benjamin/Cummings Publishing Company, Redwood City, CA.

## Grading Summary

Consult the following checklist as you prepare your report:

**Raw Data Section** (data pages from this manual) 30 points

**Cover** (format on page 78) 2 points

**Completeness** 18 points

All the data you recorded are present on the data pages.

There are no data in the Results section which cannot be found in the Raw Data section.

**Clear Association of Data with Methods** 10 points

*Every experimental condition* (even controlled variables like temperature in a protein experiment) are noted on the data pages so another person could reproduce your experiments.

**Bonus** up to 5 points

Data pages contain questions, observations, and background information in addition to data.

**Results Section** (on graph paper) 30 points:
5 for each tank experiment
10 for stocking density experiment
5 for production run

There must be a table or graph for each experiment.

**Presentation**

The format (the type of table or graph) is appropriate for the data you are presenting (consult Chapter 4).
If a table, data are summarized.
If a graph, all x-axis intervals are equal and all y-axis intervals are equal.
If a line graph, data points are shown as easily visible symbols.
If a line graph, a smooth curve is drawn through the "cloud of points."
Table or graph is neat, typewritten or computer-printed if possible.
Each table or graph is numbered and titled.

**Association of Data with Methods**

The heading of the table or caption of the graph should give the conditions of the experiment, as illustrated in Chapter 4.
Axes should be labeled and units of measurement should be given.

**Discussion/Conclusions Section** (typed or computer-printed) 40 points

**Discussion Content** 25 points

Discussion clearly states culturing recommendations.
Recommendations are justified by frequent reference to the *numbered* tables and figures of the Results.
Ambiguous results are pointed out and discussed.
Weaknesses in the experiment are discussed.
Your results are compared with those of the most profitable teams for your fish.

**Discussion Presentation** 10 points

Discussion is neatly *typed* and does not exceed the two-page limit.
Writing style is concise, clear, and free from spelling and grammar errors.

**Conclusions** 5 points

Culturing recommendations are summarized.

Conclusions agree with culturing recommendations of Discussion.

No new information is introduced in the Conclusions section.

**Bonus** up to 5 points

Discussion uses (and properly cites) relevant aquaculture material. See page 90 of this manual.

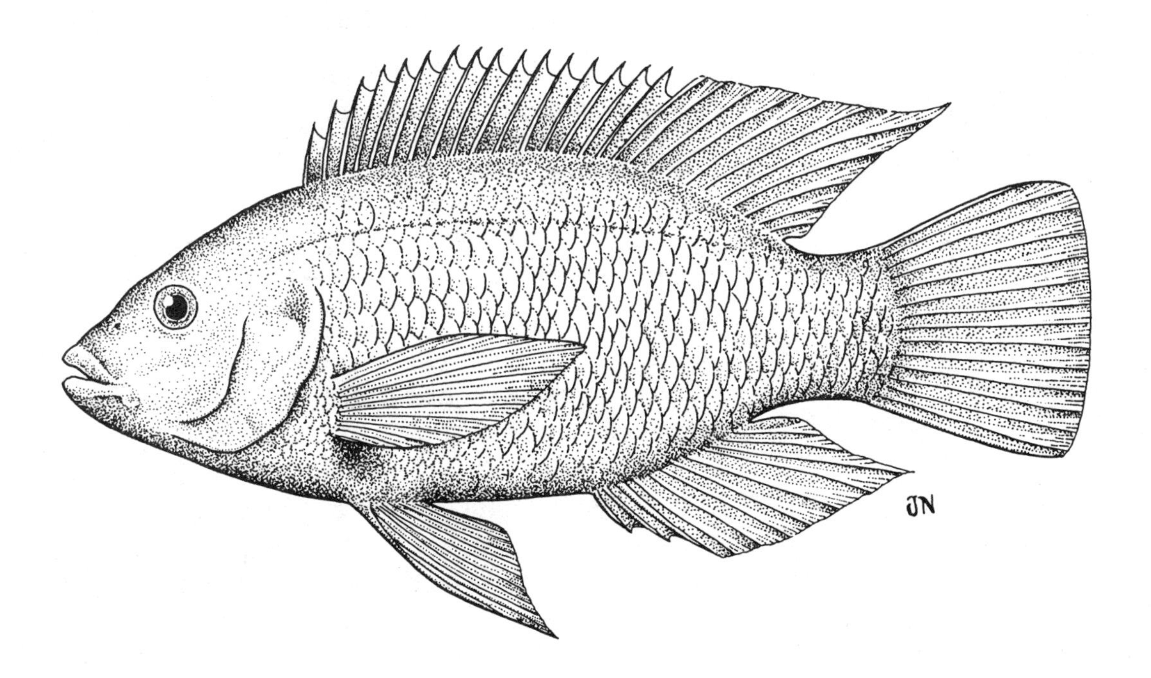


**Tilapia *(Tilapia mossambica)***

Tilapia are small warmwater fish of the Middle East and North Africa and were probably the fish in the biblical story of the loaves and fishes. At first glance, they seem to be an ideal aquaculture species. Some species are primarily herbivorous, so their feed is cheap. They are extremely tolerant of high temperatures and stagnant, polluted waters. Finally, they produce high-quality meat. However, they have one serious problem: They reproduce at very small size and produce a pond tremendously overpopulated with stunted, unsalable fish. Several African countries, Taiwan, and Indonesia are the centers of world tilapia production.

## Books on Aquaculture

Bardach, J. E., J. H. Rhyther, and W. O. McLarney. 1972. *Aquaculture: the farming and husbandry of freshwater and marine organisms.* Wiley-Interscience, New York. Big, encyclopedic, but not difficult to read. The classic book everyone quotes.

Brown, E.E. 1983. *World fish farming: cultivation and economics,* Second Edition. Avi Pub. Co., Westport, Connecticut. Country-by-country treatment.

Brown, E.E., and J. B. Gratzek. 1980. *Fish farming handbook: food, bait, tropicals and goldfish.* Avi Pub. Co., Westport, Connecticut. Has a good introduction, good catfish section.

Hickling, C.F. 1971. *Fish culture.* Faber and Faber, London. Easy to read, let-me-tell-you-a story style. How it was in Malaya in the 1950s and 1960s.

Limburg, P. R. 1980. *Farming the waters.* Beaufort Books, Inc., New York. Easy "layman's" introduction.

Meske, C. 1985. *Fish aquaculture: technology and experiments.* Pergamon Press, New York. Mostly details on carp experiments, but good introduction.

Reay, P.J. 1979. *Aquaculture.* University Park Press, Baltimore. Very short, but rather high level.

Stickney, R.R. 1979. *Principles of warmwater aquaculture.* John Wiley and Sons, New York. A discussion of general principles interspersed with illustrations taken from culture of catfish, *Tilapia,* and other warmwater fishes.

Stickney, R.R. 1986. *Culture of nonsalmonid freshwater fishes.* CRC Press, Boca Raton, Florida. A high-level overview of techniques used to raise catfish, carp, perch, bass, small baitfish, and several other kinds of fish. Each fish has an introduction which is fairly informative.

Tucker, C.S. (editor). 1985. *Channel catfish culture.* Developments in Aquaculture and Fisheries Science, Vol. 15. Elsevier, New York. A high-level catfish encyclopedia.

APPENDIX

# Fish Aquaculture

Aquaculture is the growth of captive aquatic animals for human use and consumption. Although aquaculture is not nearly as important as terrestrial agriculture in meeting world food needs, it has much promise. As long as sufficient supplies of unpolluted water are available, aquaculture can be practiced on soils which are too nutrient-poor, acidic, or impermeable for crops. It can employ water which is too salty for use in irrigation (striped mullet can grow in full-strength sea water; even "freshwater" channel catfish can tolerate 25% seawater). Also, fish are 2 to 20 times as efficient at converting feed to flesh as are cattle, pigs, and poultry. This occurs because fish are "cold-blooded" and so do not spend energy maintaining a constant body temperature, they are largely supported by water and so expend little energy on maintaining their position or moving from place to place, and their reproduction does not consume as much energy as does the reproduction of birds and mammals. Lastly, as pollution and overfishing reduce the world fish catch, aquaculture may become increasingly important as a replacement.

In 1991, aquaculture contributed about 18.2 million metric tons of meat and seaweed to the world's food supply. The most productive aquaculture countries (in rank order) were China (providing half of the world total), Japan, South Korea, the Philippines, the United States (accounting for 3% of world production), and the former USSR. The most impor tant cultured organisms in order of production were carp (25 species), oysters (11 species), mussels (9 species), an African fish called *Tilapia* (6 species), trout and salmon (9 species), catfish (9 species), clams (15 species), and shrimp (20 species). In 1991, aquaculture supplied 10% of the world's fish consumption.

## History

Aquaculture has a long history. It was practiced in ancient Egypt and Rome and in medieval Europe, and carp aquaculture advanced greatly in ancient China (the world's first known aquaculture manual was written there in 460 B.C.).

Most of these old systems were based on raising fish (especially carp) in static ponds and either relying on natural foods such as algae, insect larvae, and worms living in the bottom mud or on feeding with supplementary natural foods such as snails and small clams, spoiled food, human or animal feces, silkworm pupae or vegetable refuse from crop production. Sometimes this kind of aquaculture attained high levels of ecological sophistication. In Chinese "polyculture," several kinds of animals are raised together so that a maximum amount of food energy is converted to fish flesh. For example, pigs are raised in pens over ponds, and the pig feces and food scraps fall into the water. The large particles are eaten directly by common carp. Feces which dissolve fertilize the water and produce blooms of algae, which are eaten by silver carp. Algae which escape the silver carp cause population explosions of zooplankton, and the zooplankton are eaten by bighead carp. Finally, some nutrients may cause rapid growth of aquatic plants, which are grazed by grass carp. And to complete the circle, the common carp also eats the feces of the grass carp. For centuries, these Chinese polyculture ponds have been attaining yields of fish flesh which were only attained in the American catfish industry in the 1970s.

## Recent Developments

Since the 1950s, modern aquaculture has achieved great improvement in production by stocking organisms at far higher densities than were previously used and by feeding carefully formulated artificial diets. High stocking densities can cause the water to become foul and oxygen-depleted, and disease and fish kills due to ammonia buildup or oxygen depletion are a constant threat. There are several ways to solve the problems brought on by high stocking densities.

First, in ponds, oxygen depletion can be countered by mechanical agitation of the water surface to drive off $CO_2$ and bring in oxygen.

The "raceway solution" to the problems caused by high population densities employs a rapid flowthrough of clean water. This flushes away the fouled water and allows fish to grow and remain healthy in an enclosure where half the volume may be fish biomass. The disadvantage of raceways is that abundant supplies of high-quality water are absolutely necessary and that extremely crowded fish may all die in a few minutes if the water supply fails.

Another solution is to grow the fish in cages surrounded by large volumes of unpopulated water. Feces and uneaten food fall out of the cages and away from the fish and currents bring in a supply of fresh water. Harvest is simplified by the fact that the fish are confined in a small area.

Diet improvement has also revolutionized yields and aquaculture economics. Fish, like humans, require a balanced diet which contains adequate amounts of energy, protein, vitamins, and trace elements. The natural food of the fish (bottom organisms in the case of catfish) contains the correct mix of nutrients. However, in raceways, cages, or densely stocked ponds, the few natural food organisms are rapidly consumed and the artificial diet must supply nearly all of the nutritional demands of the fish. The cost of this diet is easily the most expensive item in the aquaculture budget, but the investment is rewarded by pond yields which were unheard of twenty years ago.

Although aquaculture has advanced rapidly, it is still much more primitive than conventional agriculture in some ways. One of these problem areas is fish genetics. Although some kinds of stock (especially carp) have been bred for centuries, we are still ignorant of the genetics which govern growth rate and yield. In other cases (striped mullet), wild stock with tremendous local variation is used. One authority concluded that with respect to our ability to control the genetics of the stock, aquaculture today is about as advanced as grain growing was in ancient Egypt. The bright side of this situation is that we may see rapid increases in yields when modern genetic engineering techniques are applied to aquacultural stock animals.

## Aquacultural Yields

The success of an aquacultural enterprise depends on converting low-cost feed into high-value fish flesh with the greatest possible efficiency. The greatest yields are attained when high biomasses of fish are continuously eating large amounts of feed. The factor which tends to limit production most decisively in this situation is oxygen depletion. With large amounts of fish, fish feces, food particles, and plant nutrients in the water, bacteria and algae (microscopic plants) attain high populations and threaten to consume all the oxygen in the water. Of course, massive fish kills result if this happens.

Yields are usually measured in kilograms of live fish per hectare per year (a kilogram, or kg, is 2.2 pounds; a hectare, or ha, is 10,000 square meters, or about 2.5 acres; 1 kg/ha is 0.88 pounds/acre). In Table 9, note how yields increase as the water becomes more fertile, the warm season becomes longer, artificial aeration is employed to counter oxygen depletion, and the water replacement rate (the percent of the enclosure's water replaced per unit time) becomes more rapid.

**Table 9.** Yields of fish (kg/hectare/yr) from several systems

| Fish | Place | Year | Yield | Comments |
|---|---|---|---|---|
| | | Static Pond Culture | | |
| All sp. | Germany | 1950s | 13 | Natural infertile lakes |
| All sp. | various | 1950s | 400 | Natural fertile lakes |
| Carp | Israel | 1971 | 100 | Unfertilized, unfed ponds |
| Carp | Israel | 1971 | 350 | Fertilized ponds |
| Carp | Israel | 1971 | 600 | Ponds fed with fish food |
| Carp | Israel | 1971 | 2680 | Ponds both fertilized and fed |
| Carp | W. Ger. | 1985 | 780 | Ponds not treated with pig manure |
| Carp | W. Ger. | 1985 | 2000 | Ponds given pig manure |
| *Tilapia* | Belgian Congo | 1949 | 7950 | Fed mill sweepings |
| Several | China | 1961 | 2800-6000 | Carp polyculture |
| Carp | Japan | 1974 | 5200 | Average, manured ponds |
| Catfish | USA | 1972 | 1500-2500 | Average yields in late 60s, using no artificial aeration |
| Catfish | USA | 1985 | 3000-6000 | Average yields in 1980s, with artificial aeration |
| | | Raceway and Cage Culture | | |
| Carp | Japan | 1974 | 120,400 | Cage culture average |
| Trout | Denmark | 1977 | 8,300 | Water replacement 12%/hr |
| Catfish | USA | 1972 | 40,000 | Water replacement 75%/hr |
| Trout | USA | 1972 | 1,360,000 | One trout/liter in tank |
| Carp | Japan | 1952 | 4,400,000 | Water replacement 4,800%/hr |

When comparing the raceway and pond yields, keep in mind that intensive raceway culture is only possible where there is an abundant water supply, and that usually the enclosures are small. The Japanese report of 4.4 million kg of carp/hectare/yr refers not to a large pond but to a 4 x 4 m enclosure which had its water volume replaced every 75 seconds. This enclosure had over 500 carp/$m^2$; a Japanese static water carp pond would be overstocked if it had **1** carp per $m^2$. Recirculating systems that do not use so much water are possible, but at a cost. For example, the recirculating culture system at Ahrensburg, Germany can culture 1500 kg of carp in 6.5 $m^3$, at approximately 600 times the normal pond density. But each 6.5 $m^3$ tank requires 46 $m^3$ of filtration, purification, settling, and sludge removal tanks attached to it.

While world catfish production is only 5% of world carp production, catfish is the most important fish in American aquaculture, and *Fish Farm* includes a simulation of catfish culture. Therefore, an overview of catfish culture in the United States is next in order.

# American Catfish Culture

## Catfish Natural History

The catfish is probably one of the most easily identifiable freshwater fish due to its "whiskers"--sensory barbels which it brushes over the bottom as it searches for food. Another notable feature is a smooth, mucus-covered skin rather than the usual scales. The mucus does an excellent job at protecting the catfish from infection, and it is reputed to speed the healing of wounds in human skin as well. Finally, catfish have sharp spines on some of their fins. The spines can cause painful wounds, and in some catfish they are actually poisonous.

Worldwide, there are more than 2,000 species of catfish, including some which swim upside down or give electric shocks, or travel overland, staying out of water for up to twelve hours. But in North America, the term *catfish* mostly refers to the family Ictaluridae, with 39 species. Of these, certainly the most important to aquaculture is the channel catfish, *Ictalurus punctatus* (Figure 19). These fish were originally found from the Gulf states and the Mississippi valley north to the Great Lakes and the prairie provinces of Canada, but now have been widely introduced elsewhere. They live in rivers, streams, and ponds and forage along the bottom at night for fish, fish eggs, frogs, crayfish, insects, and worms. In aquaculture, you will be dealing with channel catfish less than 40 cm long and 1 kg (2 pounds) in weight, but if allowed to grow, channel catfish may live for 15-20 years and can reach a length of 1.2 m and a weight of 26 kg, or almost 60 pounds. The European catfish can reach 3 m and a weight of 180 kg--400 pounds!

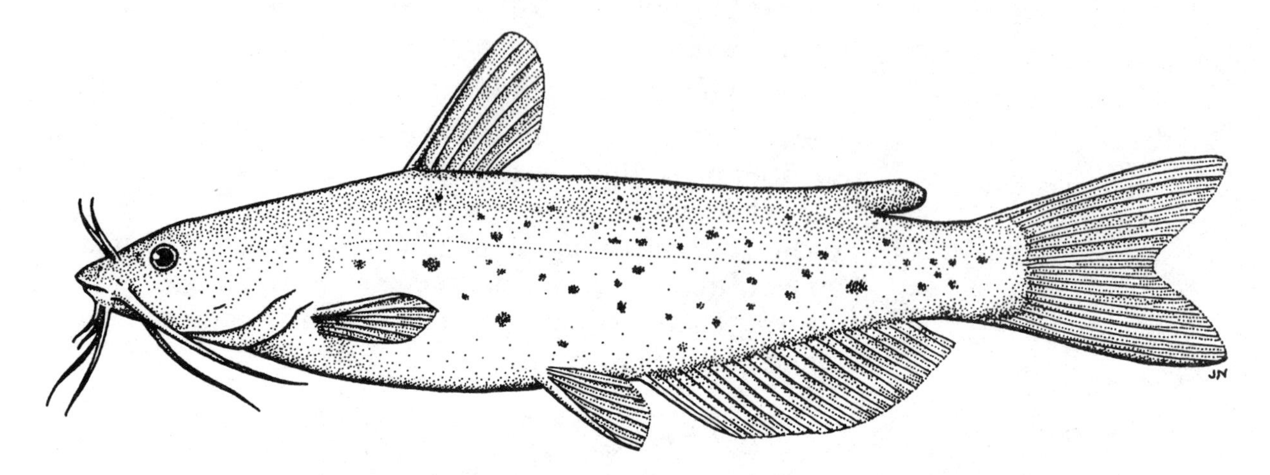

**Figure 19.** The channel catfish, *Ictalurus punctatus.*

Channel catfish have good growth rates, tolerate crowding and foul water, resist disease well, readily accept artificial diets, and have a mild flavor with wide market appeal. Also, they will not reproduce under normal pond conditions. This means that they put a maximum amount of energy into growth. At the other extreme, the African *Tilapia* is infamous for reproducing wildly and producing a pond crowded with stunted fish. For these reasons, the channel catfish is the favorite stock animal of the American catfish industry.

## The Catfish Industry

Catfish culture is the largest aquaculture industry in the United States and is one of the economic success stories of the past 20 years. Catfish culture only began on a large scale around 1963. However, growth was slow because the growers could only supply their customers with fish during the single harvest season in early winter. In the mid-1970s the multiple harvest system came into common use. In this system there are several harvests throughout the year, with only the largest fish being harvested each time. These fish are then immediately replaced with new fingerlings. Thus production and supply of fish is continuous, and the market is greatly expanded. The spectacular results for the industry are shown in Figure 20.

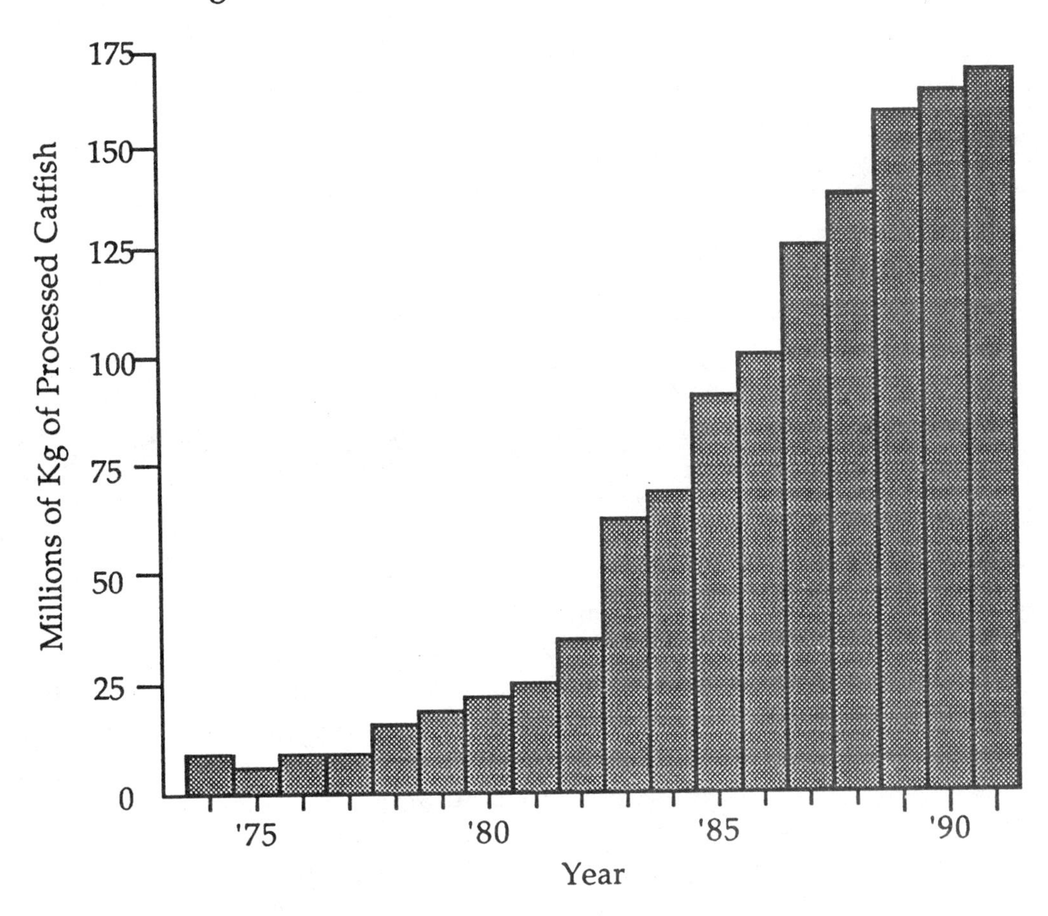


**Figure 20.** Catfish handled by U.S. processing plants, 1974-1991.

The production increase shown reflects not only an increasing number of hectares devoted to catfish farming, but also an increase in yields. In the early 1950s, yields of fertilized ponds were about 250 kg/hectare; by the late 1950s intensive research on diet had raised experimental yields to 1,500 kg/hectare; later this increased to 2,500-3,000 kg/hectare. Application of mechanical aeration raised pond yields to 3,000-6,000 kg/hectare by the late 1970s.

Typically, harvested catfish are sold live to a processing plant, and from there they are sold to retail grocery outlets, food service distributors and restaurants. A small portion of the harvest finds its way to "fee fishing" operations, where a landowner stocks a pond with fish and then charges a fee to anglers who wish to fish the pond.

Per capita consumption of catfish in the USA is 0.22 kg/yr (accounting for 3% of U.S. fish consumption). For comparison, the average American eats 22 kg of poultry each year.

The major areas for catfish production in the United States are located in Mississippi, Arkansas, Louisiana, and Alabama. The U.S. also imports catfish, mostly from Brazil.

## The Catfish Production Cycle

Catfish production is divided into a number of phases, and each may be carried out by specialized operators. In *Fish Farm,* you will only be concerned with the last ("grow-out") phase. The whole production cycle takes two years.

First, large, old "brood" fish are allowed to breed and produce eggs in spring or early summer of the first year.

The egg masses may then be moved to hatchery troughs, where they are slowly agitated by paddlewheels. These eggs hatch within a few days, and the young fish are allowed to grow for several days more before they are moved to "nursery" ponds.

The fry are reared to the "fingerling" stage (anywhere from 5 to 25 cm) by autumn of the first year. In the nursery ponds they may suffer greatly from predation by wild fish (especially sunfish) and aquatic insects.

Finally, the fingerlings are moved to "grow-out" ponds during the second year. They are stocked at a much lower population density than in the nursery ponds and fed until they reach a weight of about 500-800 g and a length of about 30 cm (1 foot). The date of stocking the fingerlings may vary in the multiple harvest systems used on real catfish farms, but in the simpler world of *Fish Farm,* you will always stock a single crop on March 21 and will harvest it on or before November 5.

FISH FARM

# Appendix A

## Installing and Running *Fish Farm* on Your Computer

*Fish Farm* is sold in either an Apple® II/IIe/II GS version, an IBM® version, or a Macintosh® version.

The Apple II version of *Fish Farm* has Apple system files on it and will start the program when it is placed in the disk drive and the computer is turned on. A disk which will do this is called a self-booting disk. But the Macintosh and IBM versions do not contain disk operating system (DOS) files and are not self-booting. The two basic ways to use the Macintosh and IBM *Fish Farm* in this case are either to install the program on a hard drive which has DOS or to transfer its files to a disk which contains DOS.

## Using the Apple® II Version

Your Apple II disk contains DOS 3.3 system files and will boot when it is placed in a disk drive and the computer is turned on. However, you should make a copy of the original disk and store the original in a safe place.

To copy the disk using DOS 3.3, boot the computer with a DOS disk in drive 1. Then type **RUN COPYA.** After the copying program has loaded, put the original *Fish Farm* disk into drive 1 and a blank disk into drive 2. Then follow the directions on the screen.

If you are using an Apple IIGS computer, the program will run faster if you set the system speed to "fast." Boot up the *Fish Farm* disk, wait until the program stops loading, and then press the "CONTROL" and open-Apple and "ESC" keys all at the same time. You will be shown a series of menus. Select "Control Panel" and then "System Speed." Exit from the menus by pressing "ESC." Once the system speed is set to fast, it will remain there even if the computer is turned off.

## Using the IBM® Version

**If you want to install *Fish Farm* on a hard drive**, create a directory named "fish" (type **md \fish** followed by the **Enter** key). Then, with the *Fish Farm* disk in drive A and the computer on the hard drive (the "C:" prompt is displayed), type **copy a:*.* \fish** followed by **Enter**. To run *Fish Farm*, get into the "fish" directory by the command **cd \fish** and then type **fish** to run the program.

**If you want to install *Fish Farm* on a bootable disk**, you must format and install DOS on a *blank* floppy disk. To do this, put the blank disk (**NOT the *Fish Farm* disk!!!**) into drive A, and format it with the command **format a: /s**. Then put the *Fish Farm* disk into drive B and transfer its files to the formatted disk with the command **copy b:*.* a:**. After the copying is complete, the disk in drive A will boot. You can run the program immediately by typing **fish** followed by the **Enter** key.

If you are running *Fish Farm* directly from a disk, place the original *Fish Farm* disk in a safe place. To ensure that you will always have a good copy of the program, do not allow the original *Fish Farm* disk to be used for routine running of the program.

## Using the Macintosh® Version

**If you want to install *Fish Farm* on a hard drive,** boot your Macintosh and insert the *Fish Farm* disk into the disk drive. Use the mouse to move the *Fish Farm* disk icon to the hard drive icon. All the files will be copied into a folder called *Fish Farm* on the hard drive. Then put your original *Fish Farm* disk in a safe place.

Thereafter, when you want to run *Fish Farm*, double-click on the *Fish Farm* folder, and then double click on the *Fish Farm* icon inside the folder.

**If you have two disk drives but no hard drive,** boot your Macintosh with a system disk in the startup drive. After the desktop screen appears, insert a blank disk (**NOT the original *Fish Farm* disk!!!**) into the second drive and follow the on-screen directions to format the blank disk. Then remove the system disk and insert the original *Fish Farm* disk into the startup drive. Click on the original *Fish Farm* disk to high-light it, and then move it to the icon for the blank disk. All the files will be moved onto the blank disk. Remove the original *Fish Farm* disk and put it in a safe place.

Thereafter, when you want to run *Fish Farm*, boot your Macintosh with a system disk in the startup drive and the *Fish Farm* copy in the second drive. Then double-click on the *Fish Farm* disk icon to open it and double-click on the *Fish Farm* icon to run *Fish Farm*.

FISH FARM

# Appendix B

## Production Run Record

## Transparency Masters

# *Fish Farm* Production Runs

Names of Team Members

____________________ ____________________

____________________ ____________________

Section _______ Fish _______

| | Conditions | | Results | |
|---|---|---|---|---|
| 1 | Aerator (mg/l) | _____ | Kg/ha harvested | _____ |
| | Groundwater (%/day) | _____ | Feed conversion ratio | _____ |
| | Feed protein (%) | _____ | Mortality rate (%) | _____ |
| | Stocking (fish/ha) | _____ | Profit or loss (+/-) | _____ |
| | Harvest time (ES or MW) | _____ | | |
| 2 | Aerator (mg/l) | _____ | Kg/ha harvested | _____ |
| | Groundwater (%/day) | _____ | Feed conversion ratio | _____ |
| | Feed protein (%) | _____ | Mortality rate (%) | _____ |
| | Stocking (fish/ha) | _____ | Profit or loss (+/-) | _____ |
| | Harvest time (ES or MW) | _____ | | |
| 3 | Aerator (mg/l) | _____ | Kg/ha harvested | _____ |
| | Groundwater (%/day) | _____ | Feed conversion ratio | _____ |
| | Feed protein (%) | _____ | Mortality rate (%) | _____ |
| | Stocking (fish/ha) | _____ | Profit or loss (+/-) | _____ |
| | Harvest time (ES or MW) | _____ | | |
| 4 | Aerator (mg/l) | _____ | Kg/ha harvested | _____ |
| | Groundwater (%/day) | _____ | Feed conversion ratio | _____ |
| | Feed protein (%) | _____ | Mortality rate (%) | _____ |
| | Stocking (fish/ha) | _____ | Profit or loss (+/-) | _____ |
| | Harvest time (ES or MW) | _____ | | |
| 5 | Aerator (mg/l) | _____ | Kg/ha harvested | _____ |
| | Groundwater (%/day) | _____ | Feed conversion ratio | _____ |
| | Feed protein (%) | _____ | Mortality rate (%) | _____ |
| | Stocking (fish/ha) | _____ | Profit or loss (+/-) | _____ |
| | Harvest time (ES or MW) | _____ | | |

**Total Profit** [ ]

**Table 1.** Temperature of outdoor ponds between March 21 and November 5 as a function of the amount of groundwater input. Input is expressed as a percentage of the pond volume which is pumped in per day.

| Inflow (%/day) | Lowest Temperature | Highest Temperature | Largest 12 Hour Fluctuation |
|---|---|---|---|
| 0% | 10.0° C | 32.7° C | 12.0° C |
| 50% | 12.1° C | 27.7° C | 8.3° C |
| 100% | 14.0° C | 24.5° C | 5.5° C |
| 150% | 16.5° C | 21.2° C | 3.0° C |

**Table 2.** Effect of groundwater input on the hectares of production ponds which can be supplied by the production unit well.

| Input (%/day) | Hectares Supplied |
|---|---|
| 0% | 200.0 |
| 1% | 190.1 |
| 5% | 38.0 |
| 50% | 3.8 |
| 100% | 1.9 |
| 190% | 1.0 |

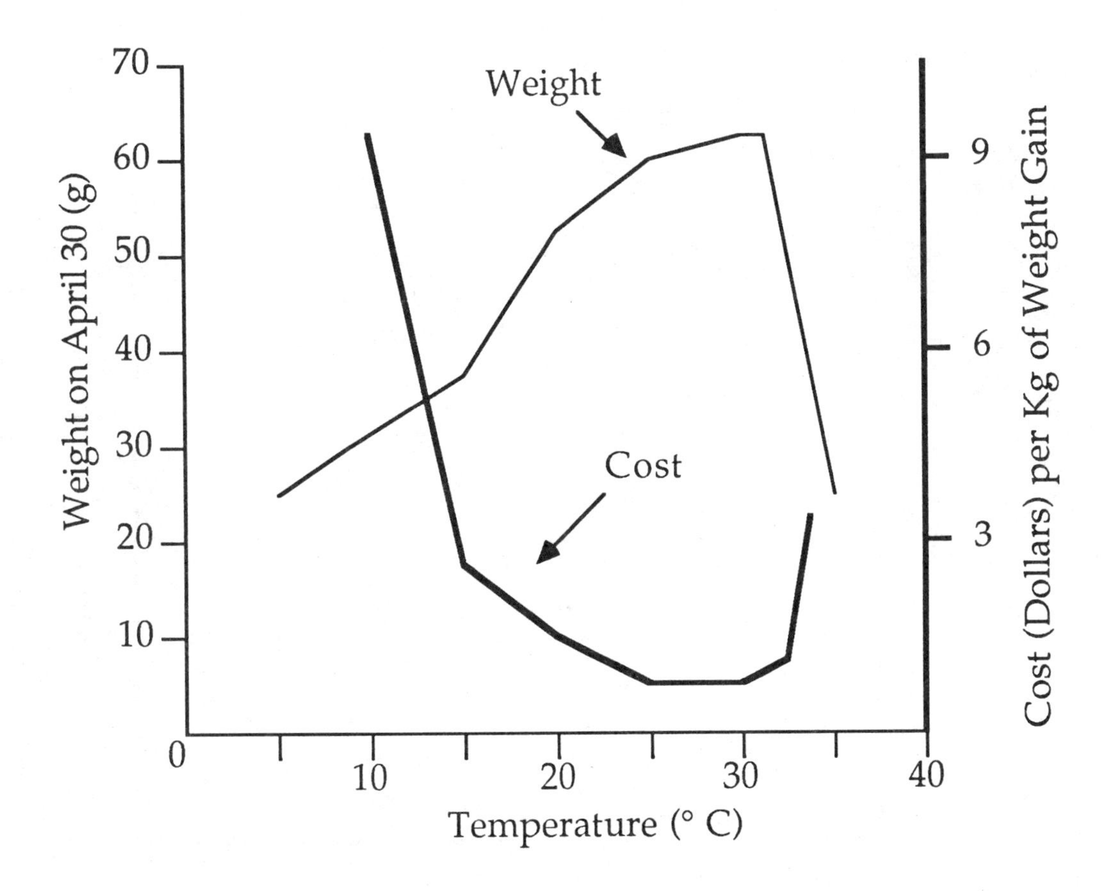


**Figure 1.** Growth of catfish as a function of temperature. Oxygen was set at 10 mg/l, and feeding at 5% of biomass/day with 50% protein feed.

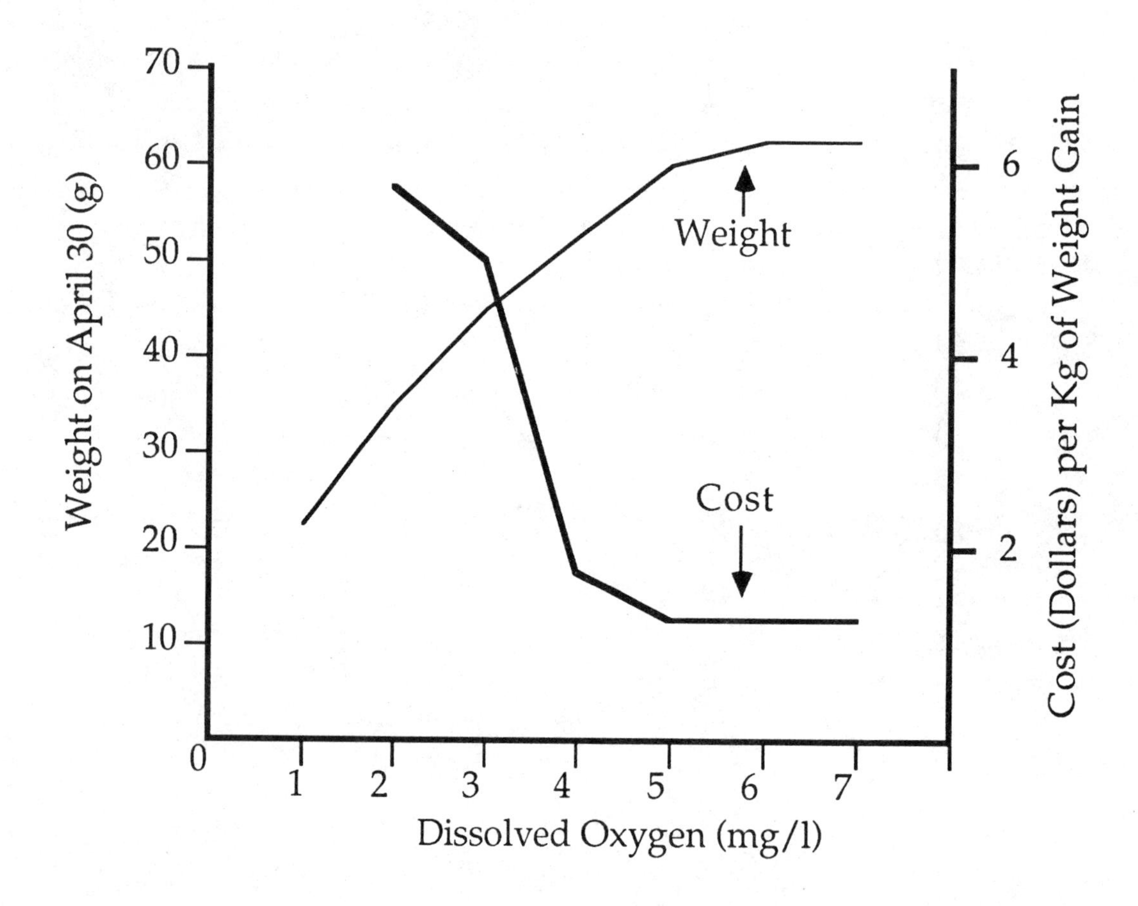


**Figure 2.** Growth of catfish as a function of dissolved oxygen. Temperature was set at 25° C, and fish were fed 5% of biomass/day with 50% protein feed.

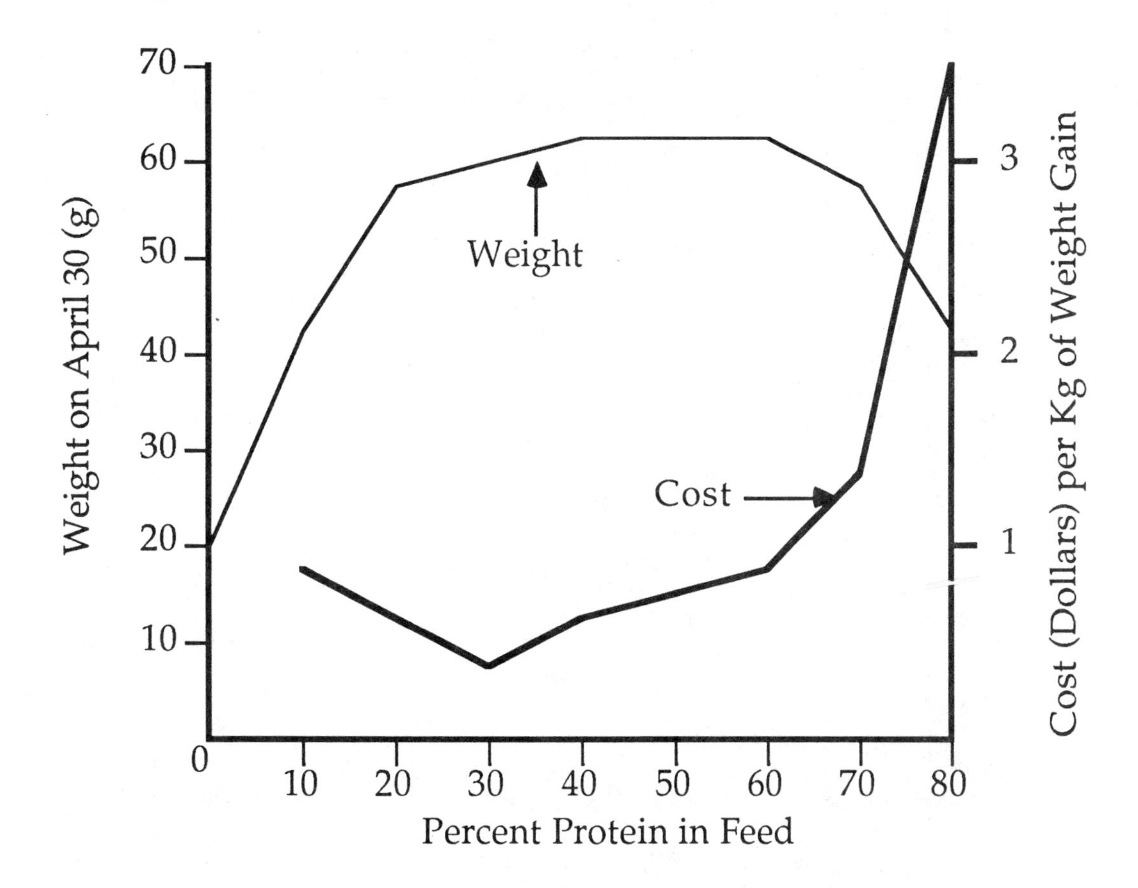


**Figure 3.** Growth of catfish as a function of protein content in the diet. Fish were grown in tanks at 25° C, 10 mg/l oxygen, and with a varying feeding rate.

## Tank Experiment Questions

1. Can the fish tolerate water temperatures of 25°-30° C without serious growth reduction? If so, it can be grown in static ponds without groundwater input.

2. How low an oxygen concentration can the fish tolerate before its growth is reduced? These data are used to determine the "trigger concentration" for the mechanical aerators.

3. What feed protein percentage produces the most economical weight gain?

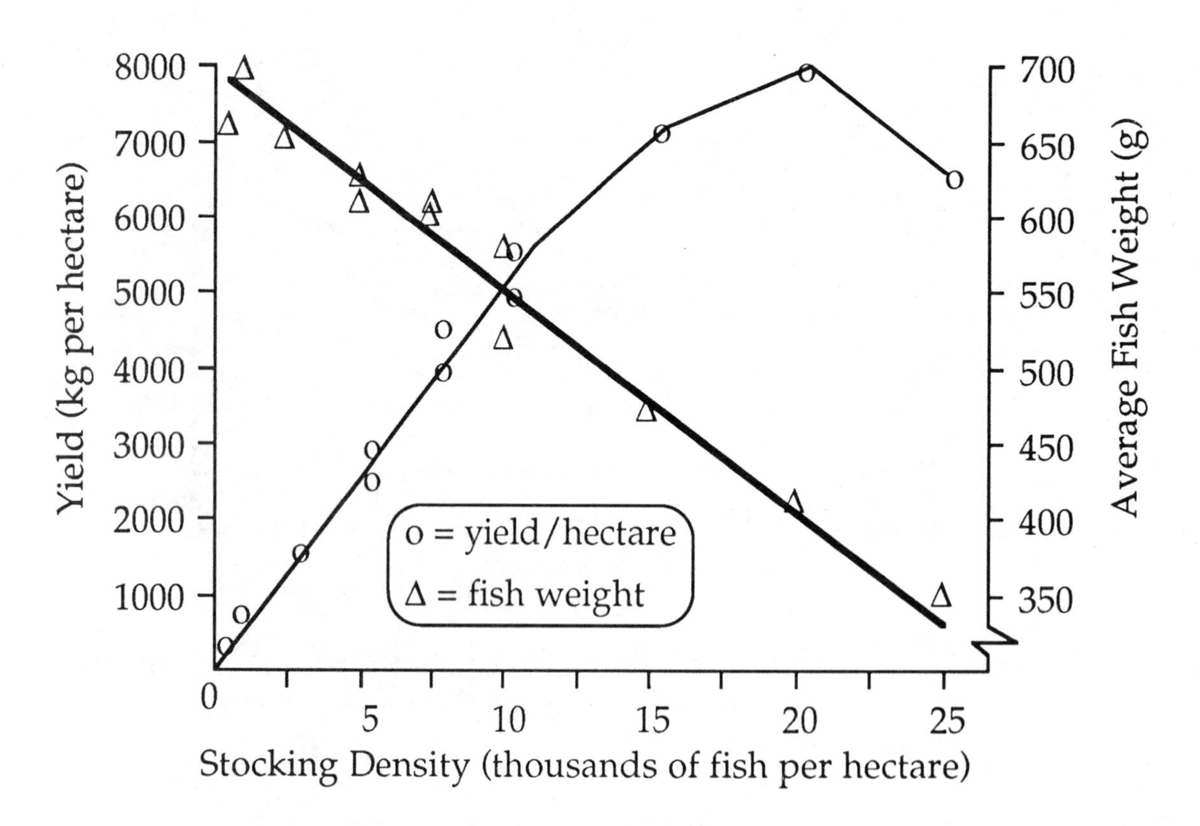


**Figure 5.** Yield and average weight of catfish from outdoor ponds. All experiments used 30% protein food, no groundwater input and aeration when dissolved oxygen reached 5 mg/l.

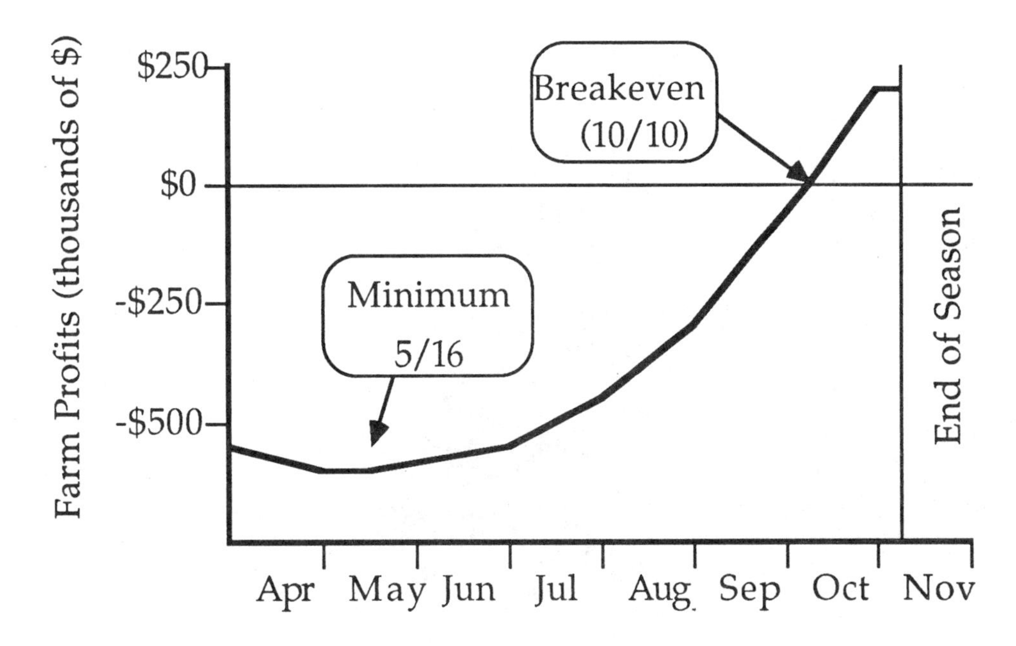


**Figure 12.** Profits for a catfish stocking density experiment. Catfish were stocked at 9,000 fish/hectare, were fed 30% protein food at a varying rate, there was no groundwater input, and emergency aeration was applied if oxygen reached 5 mg/l. Harvest was on November 5.

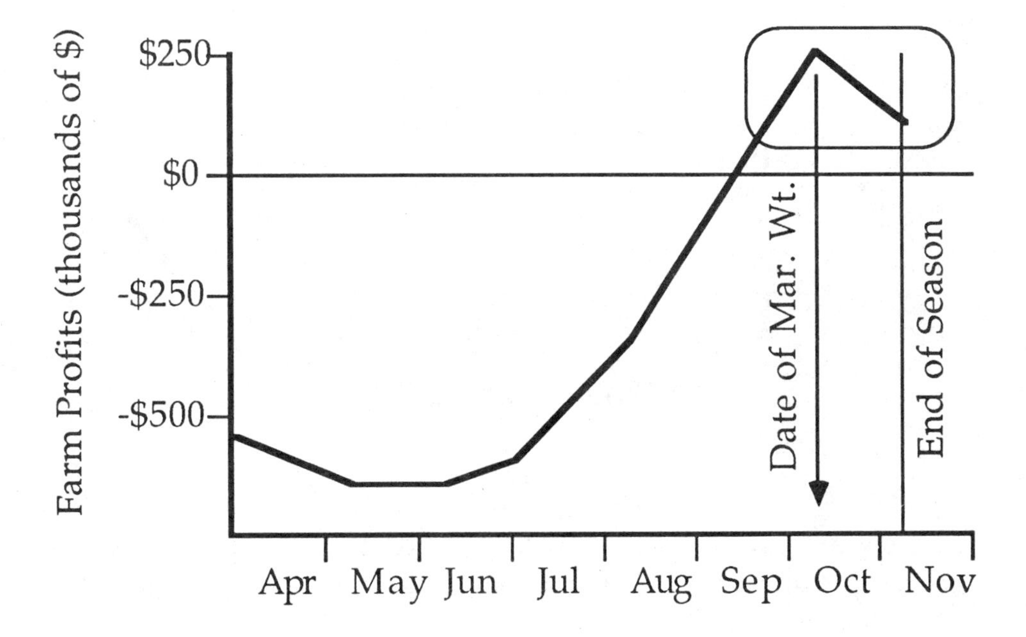


**Figure 13.** Profits for a stocking density experiment on an "unknown" fish which had a low feed conversion efficiency. Peak profits occurred on the day the fish reached marketable weight.

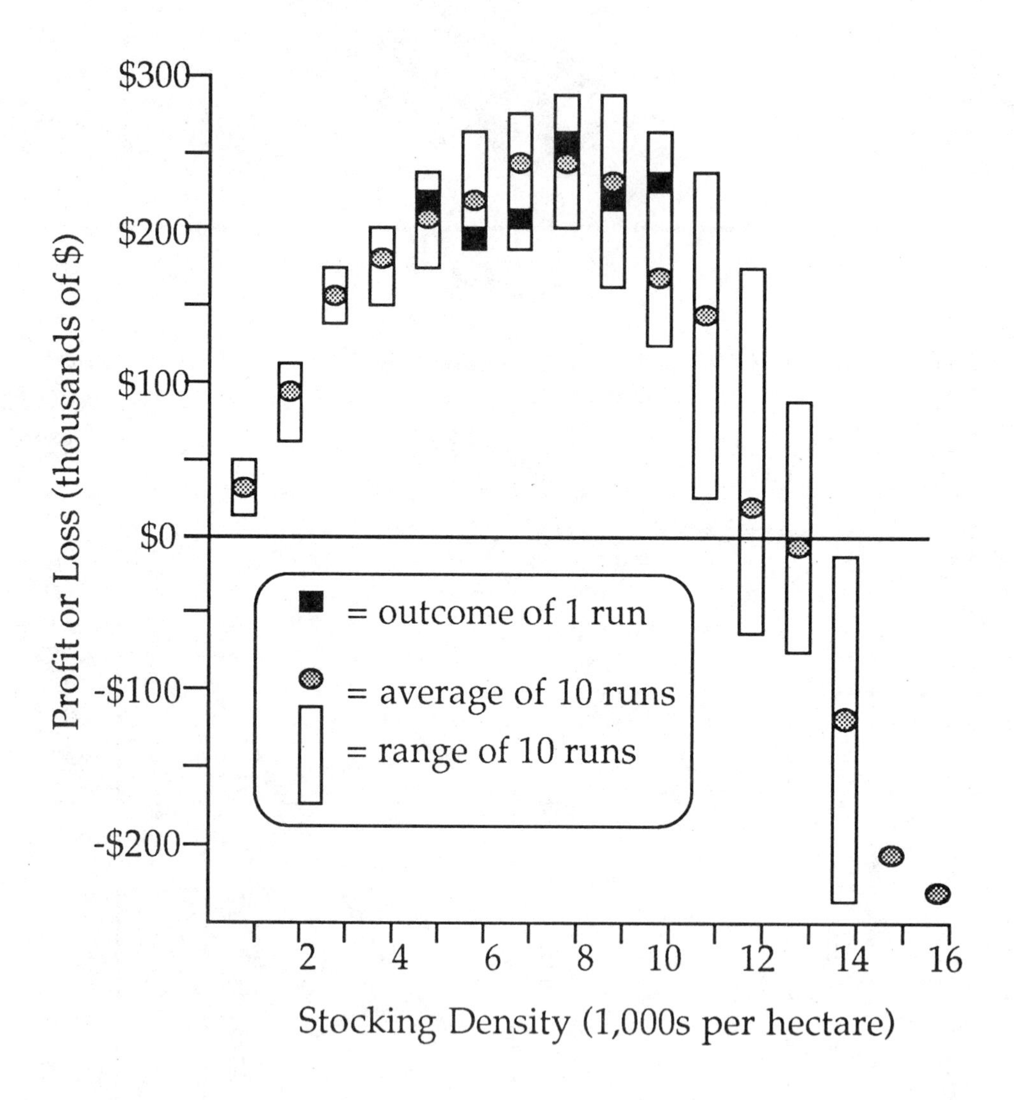


**Figure 14.** Average and range of profits from catfish stocking density experiments, computed from 10 replicate experiments per stocking density. Dark squares are results of typical single stocking density experiments.

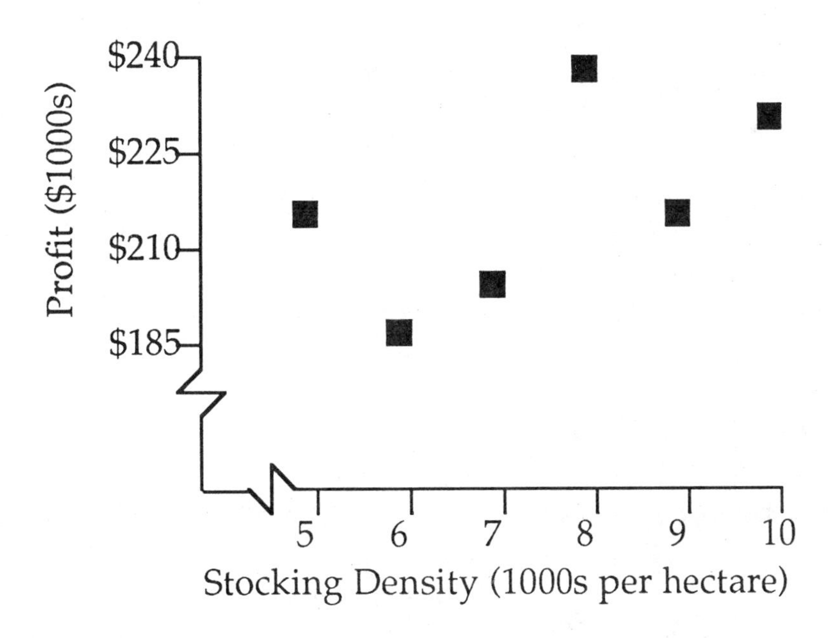


**Figure 15.** Profit results of a typical student series of catfish stocking density experiments as they would be graphed by *Fish Farm.* This experiment includes only stocking densities close to the optimum (approximately 7,500 fish per hectare).

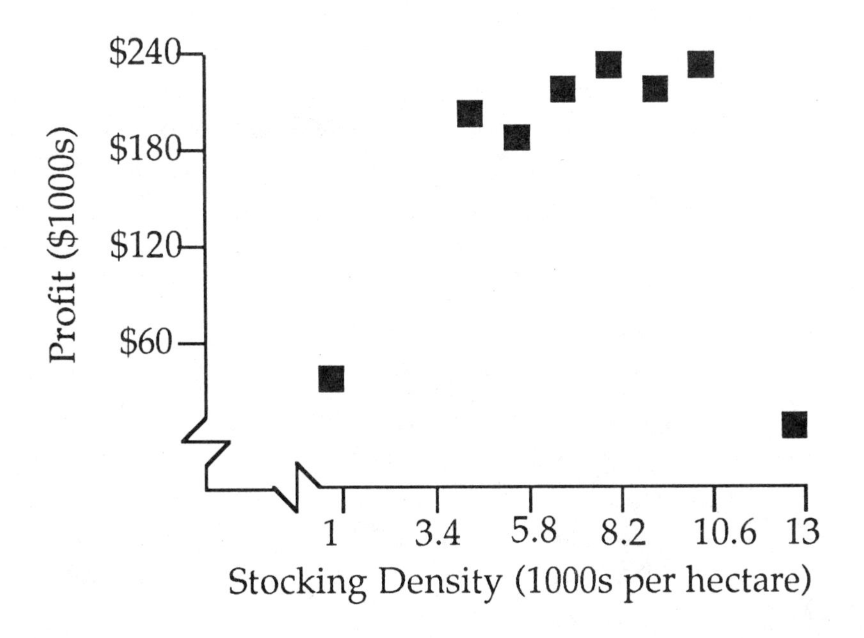


**Figure 16.** Set of student catfish stocking density experiments with inclusion of stocking densities both close to and far from the optimum.

FISH FARM

# Appendix C

## *Fish Farm's* Simulation Model

*Fish Farm* is not merely a model of fish aquaculture but rather an aquaculture model included in an ecological model of a pond ecosystem.

## Non-Fish Components of the Model

The program generates stochastically corrupted sunlight and air temperature data for Charleston, South Carolina. The mean temperatures and sunlight values are taken from NOAA data, but every time the program runs the weather is slightly different. The program iterates every 12 hours, in essence following 12 hours of noon with 12 hours of midnight.

Water temperature is a function of sunlight, air temperature, and absorption of sunlight by the water (which is mainly a function of the phytoplankton crop). The 1 m deep pond is considered to be a single mixed layer with a uniform light attenuation coefficient. The sunlight phytoplankton receive is considered to be the sunlight at a depth of 0.5 m.

Organic matter in the water is added through input of fish food and the death of bacteria, phytoplankton, benthos and fish. Organic matter is lost by bacterial decay. This decay uses up oxygen, releases ammonia (how much depends on the protein content of the feed), and causes bacterial growth. Bacteria cannot grow if oxygen is below 1 mg/l. Lost oxygen may be replaced by diffusion from the air or by photosynthesis. Ammonia is lost by diffusion to the air and by conversion to nitrate. Nitrogen is the only inorganic nutrient in the model.

Phytoplankton growth depends on light, takes up nitrate, and releases oxygen. Phytoplankton respiration releases nitrate and takes up oxygen. When the pond is overstocked the user will notice violent fluctuations in oxygen concentration. These fluctuations are caused by the repeated expansion and collapse of phytoplankton blooms over a cycle of several days. Every day, a percentage of the phytoplankton settle out on the bottom of the pond.

Pond benthos (modeled as herbivorous chironomids) eat waste fish food and phytoplankton which have settled on the bottom. If the benthos crop gets too high, the benthic organisms begin to compete for food and their growth suffers. Benthos grow as a result of eating, and respire a fraction of their biomass every 12 hours. They also have temperature and oxygen requirements for growth. There is a certain amount of benthos (in fish-proof refuges) which cannot be eaten by fish, and so the fish cannot drive the benthos extinct even with heavy feeding pressure.

Bacteria, phytoplankton and benthos all have the same temperature optimum, which slowly tracks the water temperature, but cannot keep up with it if it is changing rapidly. Their growth is depressed if the water temperature deviates too much from the optimum.

## The Fish Model

The fish model is the most elaborate part of *Fish Farm*.

Fish growth and food demand are influenced by temperature, oxygen, ammonia, disease, nutritional status, and perhaps social conflict. Each fish has an optimum temperature range (the range for catfish extends from 21° to 32°), a tolerable temperature range (for catfish, this extends from 4° to 36°), and a tolerance for temperature change within a 12 hour period (12° for catfish). Each fish also has an oxygen concentration below which it suffers growth reduction (5 mg/l for catfish) and a lethal oxygen concentration (1 mg/l for catfish), a stressful ammonia concentration (1.75 mg/l total ammonia nitrogen for catfish), a population density above which it suffers stress, and a susceptibility to disease.

Much in the fish model depends on a variable which represents stress. Temperatures outside the optimum range, rapid temperature change, oxygen below a minimum, high ammonia, weight loss, and crowding all increase stress. On the other hand, stress decays away at a constant rate. Thus if bad conditions are adding large amounts to stress, it will go up, but if additions to stress are low, stress will dwindle to zero. Stress causes fish to stop feeding and will cause them to die if it gets severe.

Disease operates synergistically with stress. If stress is high (especially when the microbial population is also high), a variable representing disease rises and adds into the stress variable. This causes stress to go up even more, which in turn causes the disease variable to go up even more. Thus, stressed fish get sick, and sick fish become even more stressed by their disease, and a catastrophic epidemic can flare up in a matter of days.

When fish die, a "Pathologist's Report" gives the percent of the *dead biomass* which is attributable to the various stressors: hot temperatures, cold temperatures, rapid temperature change, low oxygen, high ammonia, disease, and crowding. Thus if 10 fish, each weighing 27 g, die due to a cold snap, and a single 270 g fish dies due to disease, then the Pathologist's Report would show the causes of death as 50% from cold temperatures and 50% from disease (270 g lost due to each cause).

Fish food demand is highest when fish are small and stress is low (e.g., when conditions are in the optimal ranges). As fish grow larger, their food demand and respiration rate decline. When stress is high, food demand goes down and respiration rate increases, causing weight loss. Stress due to crowding is especially effective in causing increased respiration.

The only food quality attributes the program considers are whether food is artificial feed or benthic organisms, and, if it is feed, its protein content. Most fish can assimilate feed just as well as benthos, but others (e.g., Fish E and Fish U) will starve if they are not given an adequate supply of benthos. Thus their stocking densities must be low enough so that they do not crop the benthos too severely. Each fish has an optimum protein range for artificial feed (for catfish, this range extends from 30% to 60% protein). Feed with a protein content outside this range lacks appeal (relative to benthos) and will be assimilated less efficiently than feed which is inside the range. Since any feed inside the protein range will cause equivalent growth, the best feed is the one with the protein content which is at the bottom edge of the range. Feeding more protein (for example, from the middle of the range) will not cause any more growth, will increase costs, and will pollute the water with excess ammonia. If the protein content is too low, the fish will respond by eating more feed, up to a maximum amount which they can eat per day.

Conversion of eaten feed into fish flesh depends on a species-specific conversion efficiency, size, ammonia concentration, and the mix of benthos and feed eaten. Larger size and high ammonia concentrations cause the efficiency of feed conversion to decline. If a fish has a great preference for benthos and is forced to eat artificial feed, its feed conversion efficiency will be low, and it may even starve.

Appendix D shows application of these principles to a classic population ecology problem: analysis of the factors limiting fish yield.

FISH FARM
# Appendix D

## Using *Fish Farm* to Investigate Population Limitation

*Fish Farm* can be used to illustrate some classic population biology topics as well as the more applied problem of profitable aquaculture. For example, it is possible to explore population (or standing crop) limitation by trying to grow catfish at increasing densities and with various degrees of human intervention. Use harvest size (not profit) as the index of success.

If 50 catfish are stocked per hectare and not fed at all (feeding rate of 0%/day, *constant*), they will grow well and none will die. With 2000 fish, about 50% will die of starvation and there will be almost no weight gain of surviving fish. The natural benthos of the pond cannot support any more fish without mortality. The stocking density which produces the optimum yield (only 80 kg/hectare) is about 3,000 fish/hectare. Any number of fish beyond about 1,000 die rapidly, and the yield is the same no matter what the stocking density is.

The plot of individual fish weight versus stocking density is also interesting. It rapidly drops from over 800 g (at 10 fish/hectare) to about 30 g at 1,000 fish/hectare. After that, the individual fish weight remains constant. Stocking more fish only produces greater mortality, and at the end the pond produces the same 1,200 fish, each weighing 30 g.

Lack of food is the obvious problem. It is possible to remove the food barrier by feeding 30% protein feed at a varying rate (meaning the program feeds more when the fish seem hungry and cuts back when they do not seem hungry). Now the optimum yield is 4,200 kg/hectare, achieved at 10,000 fish/hectare. The fish die at higher densities, mainly from oxygen depletion and disease triggered by oxygen-related stress.

The oxygen barrier can be removed by setting the mechanical aerator to go off at 5 mg/l dissolved oxygen. Yield will increase to 7,300 kg/hectare, achieved at 28,000 fish/hectare, but disease and ammonia buildup are the new limiting factors.

Densities of 500,000 fish per hectare can be attained if a groundwater flow of 190% of the pond volume per day is used. This flow sweeps away ammonia and disease-causing bacteria and protozoans, but eventually ammonia buildup would kill the fish if stocking densities continued to increase.

The instructor might ask the students to consider the options they have--groundwater input, aeration, feeding regime, and adding antibiotic to the feed--and to hypothesize about which factors would allow the greatest yield. Then they could test their hypotheses.